Wiltshire and the Great War

Training the Empire's Soldiers

T. S. Crawford

THE CROWOOD PRESS

Also by T. S. Crawford
A History of the Umbrella (David & Charles, 1970)

In preparation:
The Canadian Army on Salisbury Plain
(Halsgrove)

First published by DPF Publishing 1999

Revised edition published in 2012 by The Crowood Press Ltd,
Ramsbury, Marlborough, Wiltshire, SN8 2HR

www.crowood.com

Proceeds from the sales of this book will be given to military charities.

© T. S. Crawford 2012

All rights reserved. No part of this publication may be reproduced or transmitted in any form or by any means, electronic or mechanical, including photocopy, recording, or any information storage and retrieval system, without permission in writing from the publishers.

British Library Cataloguing-in-Publication Data
A catalogue record for this book is available from the British Library.

ISBN 978 1 84797 355 9

The cover illustration, from a photograph taken by T. L. Fuller of Amesbury, shows soldiers believed to be those of the 59th Infantry Brigade, 20th Infantry Division, which comprised the 10th and 11th Rifle Brigade, the 10th and 11th King's Royal Rifle Corps and the 2nd Cameronians, who were based in the Lark Hill locality from April to June 1915.

Many of the illustrations in this book are taken from postcards in the author's collection. Some bear a publisher's name, others do not. In the former case, as with certain documentary material, the author has tried to identify the owners of the copyright, but eighty or ninety years after origination this has not usually proved possible. Copyright holders are invited to contact the author so acknowledgements can be made in any future edition.

Typeset by SR Nova Pvt Ltd., Bangalore, India.
Printed and bound in Spain by Graphy Cems

Contents

Introduction	5
Acknowledgements	6
Song of the Dark Ages	8
PART ONE: The War in Wiltshire: Preparation and Reality	9
1. Training for a Future War	10
2. The New Armies	28
3. Building the Wartime Camps	47
4. Supporting the Mother Country	59
5. Rail, Road and Air	75
6. Spy Scares, Censorship and Prisons	98
7. Civilians and the Army	112
8. Welfare and Women	126
9. Postcards and Postmarks	147
10. Prose and Poetry	153
11. A Slow Return to Peace	161
PART TWO: The Camps	175
Boyton and Corton	176
Bulford	177
Bustard	181
Chisledon and Draycot	183
Codford	188
Corton	194
Devizes	194
Devizes Wireless Station	196
Durrington	199
Fargo	200
Fovant	202

Hamilton	205
Heytesbury	207
Hurdcott	208
Lark Hill	211
Netheravon	216
Park House	216
Perham Down	219
Pond Farm	222
Porton	223
Rollestone	228
Sand Hill (Longbridge Deverill)	228
Sherrington	232
Sling	233
Sutton Mandeville	236
Sutton Veny	237
Tidworth	241
Tidworth Park	246
Tidworth Pennings	246
Trowbridge	250
West Down	251
Windmill Hill	252
Other Military Bases	254
Appendix: Fovant Military Badges	261
Selected Bibliography	264
Index	268

List of Maps

Bulford, Tidworth and Park House Camp	180
Codford Camp	190
Lark Hill	215
Camps near Warminster	238
Camps in Wiltshire	268

Introduction

My interest in 'military Wiltshire' began in 1960, when I was at school in the county at the time of the public campaign against the War Office's continued use of the village of Imber, which had been taken over for military training in 1943. This led to my writing a lengthy project on the Army in Wiltshire that went a little way to compensate for my limited academic achievements.

Thirty-five years later I started to collect and research old postcards on the theme. Despite having written professionally for many years, it was not at first my intention to produce a book on the subject – in any case, in 1987 N. D. G. James had written *Plain Soldiering*, which followed his *Gunners at Larkhill*. I was content to continue my own research for its own sake, until I realized there were many camps about which little was known and that no one book covered all those that existed in Wiltshire early in the twentieth century. *Plain Soldiering* is excellent in its accounts of camps actually on Salisbury Plain, but, reasonably enough given its title, does not do similar justice to those elsewhere. I was accumulating much information on all the camps, and intriguing sub-themes were emerging, notably relating to the Great War period, such as spy hysteria, Devizes Wireless Station and the change in newspapers' reporting of sexual matters, from coy allusion before 1914 to relatively graphic detail in 1918.

I thought that more could be written about the dust and mud of soldiering, showing how inappropriate to the Great War was much of the training that preceded it, when emphasis was too often placed on set-piece, textbook movements of masses of men. The result was *Wiltshire and the Great War*, which I self-published in 1999. As the centenary of the war approached, I believed that a new edition might be of interest and I was delighted when The Crowood Press agreed.

Since my initial research of the late 1990s much more information has become available, thanks mainly to the Internet. Entering the name of many camps into a search engine instantly produces a wealth of information and images. Thus I have rewritten parts of my original text to incorporate new material. This further work has impressed two points on me: the need to beware of contemporary gossip and hearsay that have come to be regarded as fact and, allied to this, the way that statistics can grow exponentially, particularly for numbers of soldiers in Wiltshire, with references to a million and even two million on Salisbury Plain. The latter figure was advanced for British soldiers alone by an anti-conscriptionist in Australia as an argument for no more of her countrymen being needed in the fighting. Conservatively totalling the capacity of various camps, airfields and barracks in Wiltshire in 1918 produces a figure of some 180,000. With the county having a pre-war population of 290,000 and with the wartime soldiers concentrated in the south of the county, the effect of this number on the civilian community in that area was immense.

The nomenclature and numbering of military units is a complex subject, with many being reorganized and retitled in 1908. The war saw a vast expansion of existing regiments and changes in designations. Specialist training units were also set up, with a sequence of titles. Simplified forms have been used, 'the 2nd Blankshire Regiment' referring to the Second Battalion of that regiment, for example.

Modern authors often provide the decimal currency equivalents of pounds, shillings and pence (£ s d), but in this book the sums of money are often too small to be meaningfully translated. Some readers may like to be reminded that one new penny represents 2.4 old ones, of which there were twelve to the shilling and 240 to the pound.

Acknowledgements

Staff and facilities at the following libraries: Aldershot; Devizes; Hampshire Local Studies Unit, Winchester; Oxford University; Prince Consort's (Aldershot); Reading; Tilehurst; Salisbury Local Studies Library; University of Reading; Wiltshire and Swindon History Centre, Chippenham; and Wiltshire Heritage Museum, Devizes.

Staff and facilities at the *Andover Advertiser*; Departments of Documents and Printed Books, Imperial War Museum; Department of Printed Books, National Army Museum; National Archives, Kew; and Salisbury and South Wiltshire Museum.

Peter Adcock for information on camps in the Fovant area; J. A. Armstrong, Dauntsey's School, West Lavington; David Bailey, Chiseldon Local History Group; Mike Barnsley for information on the Midland & South Western Junction Railway; Larry Bennett for a photograph of Devizes Wireless Station; Graham Brown, National Monuments Record, Swindon, for information on farms; Barnard Clarkson for permission to quote from Donald Clarkson's

Acknowledgements

poem, 'The Sign Post at Salisbury'; Dr Huub Driessen, Birkbeck College, London, for comments on Devizes Wireless Station; John Frost for comments on military postal services and postmarks; Jim Fuller for consenting to the reproduction of postcards published by his grandfather; Mike Harden for an image of the Fovant Military Badges; Walter Ineson for information on camps and manoeuvres in north-east Wiltshire; Marie Jones, Chemical and Biological Defence Establishment, Porton Down; Ed Kermode for permission to quote from the letters of Thomas Kermode, 8th Reinforcement, 32nd Battalion, Australian Imperial Force; Fred Lake, former librarian, Ministry of Defence, for advice on regimental journals and other sources of information; Mac McKechnie for information on many Wiltshire camps; Margaret McKenzie for information on Fovant and Hurdcott; Jean Morrison for information on the Pedrail Landship; Keith Norris for information on Porton; Ivor Slocombe for information on the alleged spy William Hacker, tribunals and local businesses; Gordon Taylor, archivist, Salvation Army.

Crown copyright material in the National Archives is reproduced by permission of the Controller of Her Majesty's Stationery Office.

Access to material deposited in the Department of Documents, Imperial War Museum, was by courtesy of the museum's trustees.

Access to Australian war diaries was by courtesy of the Australian War Memorial.

Lieutenant F. W. Paish's comments on Lark Hill are printed with the consent of A. G. C. Paish. (Lieutenant Paish's memoirs have been published by Sir Alan Peacock for private distribution under the title *War as a Temporary Occupation*.)

Corporal W. G. Beer's comments on Bulford are published with the consent of Mrs J. E. Overall.

Extracts from Victor Franklyn Smith's diary are printed by permission of John Lionel Victor Smith.

Members of the Great War Forum (www.1914-1918.invisionzone.com) for much expert guidance on points of detail.

Song of the Dark Ages

by Francis Brett Young

We digged our trenches on the down
    Beside old barrows, and the wet
White chalk we shovelled from below;
It lay like drifts of thawing snow
    On parados and parapet:

Until a pick neither struck flint
    Nor split the yielding chalky soil,
But only calcined human bone:
Poor relic of that Age of Stone
    Whose ossuary was our spoil.

Home we marched singing in the rain
    And all the while, beneath our song,
I mused how many springs should wane
And still our trenches scar the plain:
    The monument of an old wrong.

But then, I thought, the fair green sod
    Will wholly cover that white stain,
And soften, as it clothes the face
Of those old barrows, every trace
    Of violence to the patient plain.

And careless people, passing by,
    Will speak of both in casual tone:
Saying: 'You see the toil they made:
The age of iron, pick and spade,
    Here jostles with the Age of Stone.'

Yet either from that happier race
    Will merit but a passing glance;
And they will leave us both alone:
Poor savages who wrought in stone –
    Poor savages who fought in France.

After joining the Royal Army Medical Corps in 1915, Francis Brett Young was based at Sling Camp, Bulford, and in 1918 became registrar at Tidworth Military Hospital. 'Song of the Dark Ages' appeared in his Poems 1916–1918 (Collins, 1919) and is reproduced here by courtesy of David Higham Associates.

PART ONE: THE WAR IN WILTSHIRE: PREPARATION AND REALITY

1
Training for a Future War

Early Military Activity

Much of this book is about Salisbury Plain, the 'Great Plain', that bleak, undulating plateau dominating southern Wiltshire, crossed by few roads and populated only in its valleys. The seventeenth-century antiquarian John Aubrey, himself a Wiltshireman, commented that 'the traveller ... wants here only a variety of objects to make his journey lesse tedious; for here ... is not a tree, or rarely a bush to shelter one from a shower'. On a warm, sunny day the Plain is exhilarating and offers good views, notably from its northern escarpment. On a wet day all of it is as grey as Stonehenge: the sky is ash-grey, the turf is green-grey, the chalk tracks are white-grey. It is wretched. Little wonder that heavily laden soldiers marching over it were wont to overestimate the miles travelled.

Ancient man favoured the Plain for its high ground, drier than the river valleys and providing vantage points from which to detect approaching enemies. (He was more astute than his twentieth-century descendants, who unwisely chose several sites close to rivers for Great War army camps and suffered the consequences during wet weather.) There he built earthwork castles or camps such as those at Bratton, Casterley and Yarnbury, buried his dead, and left the enigmas of Stonehenge and other monuments. The Romans chose the Wiltshire downs for their villages but the valleys for their villas. These were abandoned after they quit Britain, leaving their Saxon successors to establish the lowland villages that exist today. The high Plain was left to nature.

Enclosure acts of the eighteenth and nineteenth centuries facilitated landscape planning and the planting of clumps and shelter-belts of beeches, but the Plain remained one of the largest tracts of desolate, isolated land in an increasingly populated southern England. In the early nineteenth century it was the preserve of sheep, with an estimated 500,000 of the Wiltshire Horn breed alone grazing there. By 1850 most had disappeared, leaving the Plain ready for new occupants – soldiers. In July 1855 the *Warminster Miscellany* referred to a 'proposal for a general and permanent military encampment on Salisbury Plain', but nothing more is recorded. (A correspondent to *The Times* in 1866 was to suggest 'a general [military] camp ... once a year for a week or more, on ... say, Salisbury Plain'.) Until two years previously there had been no manoeuvres anywhere in the country, making Britain the only major European power not to have annual large-scale exercises. A temporary camp created at Aldershot during the Crimean War became permanent and evolved into the home of the British Army. This may have been conveniently close to London, but its open areas were not particularly spacious and there were too many nearby towns and villages to allow free movement of large numbers of troops. In fact, manoeuvres of any size were not held there until 1871.

Their success encouraged a repetition the next year but in a wider, less-populated, area in Wiltshire and Dorset bounded by Ringwood, Longbridge Deverill, Little Cheverell, Woodborough, Grateley and Salisbury, with the Plain in its centre. To facilitate the movement of troops over private land a special Act of Parliament was passed, as had been needed for the Aldershot exercises, but permission still had to be sought from owners and farmers. Thirty-two temporary camping-grounds were laid out, including Fighleldean Field, close to the future site of Netheravon Airfield, and at or near Codford, Durrington Field and Bulford, all destined to have large Army camps built on them in the next century. Beacon Hill, near Bulford, was the scene of a concluding review of 30,000 troops by the Prince of Wales.

By the early 1890s the British Army was finding the area around Aldershot too limited, with any extension being inhibited by housing development that was also pushing up the price of land. In 1892 the Military Lands Act enabled the Government to acquire ground for military purposes. The following year manoeuvres were held in Berkshire and a small part of Wiltshire to the east of the Chiseldon–Ogbourne road, with a water supply for the troops being arranged near Manor House, Liddington. Three years later the War Department considered purchasing much of the Marlborough and western Berkshire Downs.* Then in January 1897 the Government announced that it intended to spend £450,000 on acquiring 40,000 acres of Salisbury Plain for manoeuvres, though at that time there were no plans to build permanent barracks. The first purchase was made on 3 August 1897 when land near West Lavington was acquired, and by the end of the year more than 13,600 acres had been bought. Though there was a need for cavalry to manoeuvre on open grassland, the area was also judged suitable for deploying large bodies of infantry and horse-drawn artillery. Its spaciousness would enable the creation of ranges, both for artillery and rifles; in 1897 there was none in England long enough to provide effective practice in firing the Lee-Enfield rifle introduced two years earlier.

---

*The terms 'War Office' and 'War Department' related to the same institution and both forms were in use.

By March 1900, 42,000 acres of the Plain had been bought for £560,000. Owners of some 26,000 acres sold their land voluntarily but others did so under compulsory purchase regulations. The area consisted of the Plain's central and eastern sections, divided by the Avon Valley and its various villages. Initially the area north of Amesbury was allocated for artillery practice, with a site for a tented camp laid down at Lark Hill,* but the range area itself was progressively enlarged. The eastern section was used mainly for manoeuvres, cavalry exercises and rifle ranges. Over the years other land was bought, including in 1910–12 areas that extended the West Down artillery ranges westward, and in 1927 much of the western Plain. This portion, between Westbury, Warminster, Heytesbury, Tilshead and Lavington, was sometimes used by the Army before then, notably in 1910 and during the Great War.

A Military Manoeuvres Act passed in August 1897 allowed for the closure of roads (subject to notice and the consent of Justices of the Peace), compensation for damage to property and crops, and penalties for wilful and unlawful obstruction of manoeuvres. The Military Works Act of 1898 authorized £1.6 million to be spent on accommodating seven infantry and six field artillery battalions in barracks to be erected at Bulford and Tidworth.

The barracks were duly built and a number of sites established for tented camps in summer, where the only permanent structures might be a derelict farm building, an open-sided cookhouse or a water-pumping station. The garrisons of the two new barracks provided a permanent military presence, but each year dozens of regiments of Regular and part-time soldiers descended on the Plain, usually by train, for summer camps, exercises and manoeuvres. There would be little accurate anticipation of the sort of soldiering that was to be experienced in the Great War, but many of the technological developments of the new century would be tested on the Plain for their military usefulness.

Yearly Manoeuvres

The 1897 military exercises on Salisbury Plain were modest and not facilitated by the fact that as yet the Army had manoeuvring rights over only a small portion of land. The first arrivals on the Plain were a Royal Engineers company, which marked out a camping-ground near Bulford and arranged a water supply pumped through an iron pipe from Nine Mile Stream into troughs and tanks. Early in July the 4th Cavalry Brigade appeared, with the 1st Dragoon Guards riding from Norwich, followed by the Scots Greys from Hounslow and the 3rd Dragoon Guards from Woolwich.

---

*'Lark Hill' (occasionally 'Lark's Hill or even 'Hill of Larks') was the early form of the camp's name, but soon came to be rendered as one word.

Training for a Future War

The *Morning Post* thought the War Office had been 'unbusinesslike' in not approaching landowners beforehand about using their land – negotiations appeared to be still going on during the exercises – noting that tenant farmers and sheep-grazers could not be dispossessed at a moment's notice. Shortly after the exercises started, one report said the area designated for them was 5 miles long by 1 mile wide, narrowing at one point to 300yd where a Mr Knowles refused to allow troops on his land. The many rabbit holes on Bulford Down made it impractical for horses and filling in these was to become a tiresome but necessary chore. Nevertheless, the cavalry was able to drill 3 miles to the north, on Haxton Down, and some at least of the last-minute negotiations with landowners appear to have succeeded, for exercises took place either side of the Amesbury–Andover road, with 'invaders' from Bulford attempting to collect imaginary supplies from Andover.

Training in earnest started on Salisbury Plain on 1 July 1898 with cavalry manoeuvres. The new military area was generally well liked, though some thought it unfortunate that it was divided by the Avon Valley, whose string of villages and hamlets restricted the movement of troops from one part to the other. In gaps in this string the Army established three river crossings; these were between Netheravon and Figheldean (pronounced 'Filedean'), near Syrencot House and at Fifield. Though the Government was now the major landlord, the original tenants preserved rights until the following year, further limiting troop movements. One tale, perhaps apocryphal, was of a brigade commander who guarded his front with an out-of-bounds field of crops and his left flank with an untouchable barbed-wire fence! (By 1901 this problem had eased with the introduction of three schedules of Army land leased to farmers. Tenants of Schedule I land were allowed to farm it, with the Army paying compensation for any damage; Schedule II land had to be maintained as grassland; and Schedule III land could be farmed, but with no compensation for damage by military activities.)

The Military Manoeuvres Act did not come into force until 15 August 1898, leaving little time to prepare for exercises due to start in Dorset and south Wiltshire on 1 September and ironically being held mostly away from the land purchased by the War Department. So the Army was not able to use its new compulsory powers and had to hire areas for camps through negotiation. The Royal Engineers spent several weeks laying miles of 4in steel pipe, digging wells and installing waterproof sheets to contain water, as well as installing 240 miles of telegraph wire between the camping-grounds. A Red Army mustered 25,000 men at Salisbury, its intention being to intercept the invading Blue Army, which was deemed to have 'landed' at Poole (in fact having assembled at Wareham) and was trying to link up with a force from the Bristol area. Successive days' exercises took the armies to the Fovant and Dinton areas, into Dorset then to the Wylye Valley and the Woodford locality.

The 50,000 combatants fought the 'battles' of Melbury Down and Stonehenge, the latter being more focused on the ancient earthworks of Yarnbury Castle (which lent themselves to defence and attack) than Stonehenge itself, almost 6 miles away. As a finale, 50,000 troops, 211 guns and 1,000 horses assembled at Porton Down and marched to Boscombe Down to be reviewed; it took two and a half hours for them to pass the saluting base, in front of 80,000 spectators.

There were many assessments of the new training area. Though its value was recognized, not least being large enough to allow Militia, Yeomanry and Volunteers to join with Regular troops, there was a feeling that the Army had been too ambitious with the scale of the manoeuvres.* The local rail network, designed only for civilian traffic, had yet to be improved. The roads were too poorly surfaced for military traffic (damage to them causing complaints from local councils), the Army had had to rely on civilians for extra vehicles, drivers and supplies, and there was insufficient water. The manoeuvres themselves were judged unrealistic because they were planned so that the opposing armies would find themselves at predetermined camping-grounds each night, there were ceasefires for lunch at 1pm, and supplies were provided at convenient spots (civilian drivers permitting), where they were sometimes shared by both forces. Some units, notably Militia, merely marched and never fired their rifles. All troops had to avoid woodland, sometimes diverting a mile around a 60yd strip of timber; this lent a particular artificiality in well-wooded areas where troops would have massed in actual warfare. It was also pointed out that during their manoeuvres the Germans could be billeted in private houses, whereas British soldiers had to make do with tents, which not only were not so comfortable but had to be transported from place to place.

In 1899, though much was being done to improve the Plain's roads and water supplies, the Army still had a problem with transport, lacking sufficient vehicles of its own to supply very large troop concentrations and with most improvements to the railways still at the planning stage. Therefore the authorities decided that until the infrastructure had been improved the Plain would be used to train small forces that could be mobilized fully equipped. The demands of the Boer War also inhibited large-scale deployments at home. But Militia, Yeomanry and Volunteers took their annual training on the Plain,

---

*Militia units had their origins as local forces drawn from civilians. Recruits were engaged for six years and given from three to six months' initial training, followed by from twenty-seven to fifty-six days' annual training in the case of infantry. They had permanent adjutants, quartermasters and non-commissioned officers and were liable for service only in the United Kingdom. Yeomanry were mounted troops, recruited for home defence on a county basis from farmers and yeomen and officered by country gentlemen, all providing their own horses. Until 1901 they had no Regulars attached to provide local administration. Many Volunteer Battalions evolved from Rifle Volunteer Corps, formed by Lord Lieutenants from 1859 onward.

Training for a Future War

living under canvas on camping-grounds such as Park House, Perham Down and East and West Down. *The Times* was certain about the Plain's value to Militia, noting that fifteen battalions had greatly benefited from spending a month brigaded with Regular troops.

The death of a soldier in Hampshire during a heatwave in the early summer of 1900 led to instructions being issued for the health and safety of troops on Salisbury Plain. (There would be continuing concern about part-time soldiers, who were not much used to the open air, spending two weeks out of doors at summer camp.) It was stipulated that, in hot weather, parades and musketry practice should take place in the early morning and evening; that tea, coffee or cocoa be issued to all ranks early in the day; and that men in the field for a considerable time should take with them bread, cheese and full water bottles. At West Down Camp the daily drinking allowance was one and a half gallons of water per man. Water carts and ambulances also had to accompany the troops, who were not allowed to wear serge tunics on hot mornings but paraded in their grey flannel shirts. The slouch hat was also introduced from overseas as a cooler alternative to traditional Home Service headgear, which was often heavy and hot.

Inevitably there was a criminal element among the soldiers, with petty theft, burglary and setting hayricks on fire being the sorts of cases heard by local magistrates. To reduce the problem, troops needed special passes to go into villages, which were patrolled by pickets. Nevertheless, one July weekend, eighteen Militia members ended up in Andover Police Station for 'various breaches of discipline'. And inter-regimental rivalry sometimes turned nasty, as at Bulford in early July, when several men were badly injured after jeering and chaffing between the Royal Irish and Gloucestershire regiments progressed to blows.

Men of the Oxfordshire Light Infantry on manoeuvres near Park House Camp in 1902.

Training for a Future War

It was announced in May 1903 that all troops based on Salisbury Plain would take part in weekly tactical exercises to practise positioning for assaults. Particular attention was to be paid to concealment of guns and men, whether in defence or attack, with, wherever possible, Royal Engineers detachments with entrenching tools being attached to either side. In reality there were restrictions on cutting into the Plain's turf, and not until the Great War was much practice given in constructing military earthworks.

The *Daily Express* of 13 August noted that the Plain 'is an ideal country for cavalry and artillery but not so good for infantry, for there is little cover, and foot soldiers are here invariably exposed to long-distance firing'. It quoted an officer as saying 'True, we could lose Aldershot on Salisbury Plain, and still have a wide field for manoeuvring purposes, but we could do so much more if we had more land.'

Perhaps it was this realization that led to the holding of manoeuvres in September in the much wider area of northern Wiltshire, western Hampshire, southern Oxfordshire and western Berkshire. At one time the exercises were threatened by a late harvest (though it was suggested that farmers, who were anticipating a poor yield, might have welcomed troops showing 'more than ordinary disregard for standing crops' because of the compensation available). A special system of wireless telegraphy was employed, with low receiver poles and a very short spark to limit the range of transmissions and prevent interference from 'enemy' transmitters. Balloons were used to direct the different phases of the 'battles' – one ball hung from the balloon meant a temporary cessation, two balls a continuation and three termination for the day.

Searchlights, which it was thought would be useful in night operations, notably in defensive situations, had excited the enthusiasm of the Royal Engineers for several years but they were cumbersome, being mounted on wheels 5ft in diameter. They also required a powerful and heavy generator and their reflectors were vulnerable to enemy bullets. (Wiltshire councillors were concerned they would frighten sheep.)

At this time military thinking revolved very much around an invasion of England, public nervousness of which was reflected in – and increased by – such popular books as Erskine Childers' *The Riddle of the Sands* (1903), in which two yachtsmen discover German preparations to invade England, and William Le Queux's predictive *The Invasion of 1910*, which was serialized in the *Daily Mail* in 1906. In 1903 *The Times* argued that there was an unreality about using large, open spaces such as Salisbury Plain and the Berkshire Downs to practise counter-invasion measures. Manoeuvres there did not simulate attempts to repel an invasion force on the coast and the troops were unable to use 'the close country of England [which] ... is a natural defence'.

Writing in the *United Service Magazine* in 1905, 'Foresight' was dubious about the value of special manoeuvre areas, pointing out that at Aldershot 'still we go with the same old training year after year, day after day ... and which the taught soon know by heart, and of which the teacher knows every bush'. He asked if the errors were going to be continued and accentuated by doing the same thing over again on Salisbury Plain.

Reporting on the 1906 training on the Plain, Lieutenant-General Sir Ian Hamilton criticized the 'partial, desultory, and disappointing' use of machine guns at a time when they were assuming more importance, as in the recent Russo-Japanese war, where Japanese infantry suffered from the Russian guns. A few years later, the weapon's potency would be fully utilized on the battlefields of France (and to a lesser extent in the Gallipoli campaign, for whose failure Sir Ian was made the scapegoat). The intervening period was to see increasing international tensions, notably between Germany and France. Britain was nervous of Germany's growth as a naval power and of its sabre-rattling aimed at France, Russia and the British Empire. Russo-British relations also became strained, not least over conflicting interests in Persia, just as Austria-Hungary, Germany and Russia were at odds in the Balkans, where eventually the Great War would be sparked off by the assassination of Austrian Archduke Franz Ferdinand.

In 1907 the Army's combined Southern and Eastern Commands exercised 20,000 troops in southern Wiltshire. Perhaps conscious of the criticism that very few exercises were on the coast, the War Department decreed that the

The Honourable Artillery Company, founded in 1537, was a regular visitor to the Bulford locality before the war, first as a Volunteer unit, then, from 1908, as part of the Territorial Force.

land east of the River Avon was to be regarded as the sea, with safe anchorages at Figheldean and south of Wilton, near Salisbury. Even so, the exercises failed to give any practice in preventing an invasion, for they started after the invaders – 'Blue Force' – had 'landed' and consolidated at Marlborough. Matters were not helped by food supplies and transport not being where they were expected to be, and an overnight 'armistice' had to be extended until the matter was rectified. When the 8th Hussars, part of the Blue Force, captured Sir Frederick Stopford, the defending Red Force commander, and Brigadier-General Samuel Lomax, these two worthies were released so they could ensure effective deployment of their troops!

The year 1908 saw the introduction of many Army reforms proposed by the Secretary of State for War, Richard Haldane, including the creation of a part-time Territorial Force embracing the Yeomanry and Volunteers. Most of the former Yeomanry and Volunteer units continued under new designations with the same regiments to which they had been attached previously. From 1908 usually the 1st and 2nd battalions of a regiment were composed of Regular troops (one often serving overseas), the 3rd was the Special Reserve (formerly the Militia), and the 4th and subsequent battalions were drawn from Territorials. The Haldane review also led to a switch in military thinking away from an invasion of England towards a Continental war, with plans for an expeditionary force to cross the Channel in the event of hostilities. An Officers' Training Corps was also introduced, with members drawn from public schools and universities and intended to provide candidates for officers' commissions.

Units of the new Territorial Force mustered on the Plain in August. The South Midland Division's performance was watched with particular interest because it was the first exercise involving a complete Territorial division, comprising men trained by their own officers, who would lead them in wartime. The Territorials, who also included the 1st and 2nd London Divisions, performed creditably, their achievements including building a 50ft bridge on four trestles over the River Avon at Fittleton.

But there was concern about the disappointing numbers of part-time soldiers. The 6th Light Brigade, with an establishment (or allocated number of posts) of 3,840 men, excluding officers, had only 1,287 on its strength; of these, 1,043 were in camp on Salisbury Plain, but 45 per cent of these attended for only eight days, rather than a full fortnight, as did 307 of the 14th London Regiment (the 'London Scottish'), which had an establishment of 960, a strength of 675 and only 491 at camp. Much of the blame was laid on employers' reluctance to release workers for longer.

Wireless telegraphy proved useful but only sometimes, as its reliability was subject to atmospheric conditions. In theory, one mobile station could communicate to another that was 50 miles away, though a disadvantage was that the operators needed to work from a stationary base. A suggestion that two telegraphy stations, one operating, the other on the move, should be attached to the main body of men was rejected because extra transport would be needed, as would a protective escort for the mobile station.

In late September 1909 major exercises involving 50,000 troops took place in an area bounded by Cheltenham, Northampton, Oxford and Salisbury, the last city being the headquarters of the Blue Army, whose territory extended to the River Thames, close to which most of the action took place. Observers from nineteen countries attended. (A tradition in Highworth had it that Germany's Kaiser was present, but this would certainly have been a case of mistaken identity or village gossip.)

At the end of August, a two-day cavalry reconnaissance exercise was held in eastern Wiltshire and western Berkshire. Wireless telegraphy units were attached to the cavalry, but their equipment was still too cumbersome for effective use in the field, so experiments were being made with acetylene lamps that could transmit messages in daylight, without needing the sunshine as did heliographs. *The Times* thought motorcycles a better way of communicating, though this was not borne out by message-sending trials in September, involving a motorcyclist, a heliograph – and a horse-rider, who won.

That year the Army Act was amended to allow billeting of troops during an emergency. So, during the cavalry divisional training in Dorset and Wiltshire in August 1910, the Army experimented with a voluntary scheme, inviting farmers to provide overnight accommodation for a payment of 1s for each man and horse. The men usually slept in barns, as did officers unless they were invited to stay in the farmhouse. Major R. L. Mullens drove 2,976 miles in his own car (without receiving any form of mileage allowance) to make all the arrangements and found little trouble in securing sufficient offers.

Observing activities on the Plain in early August, a correspondent for *The Times* repeated the feeling that they were somewhat artificial:

> The final issue at which the position was decided happily coincided with the place where the march past was to take place ... it cannot be said that these rapid operations bore much resemblance to what might be expected in war in similar circumstances, and the unreliability of such a morning's work, in the opinion of many, exercised a dispiriting influence on the minds of those who take their training [seriously].

In mid-September 1910 some 50,000 troops under the control of the commander-in-chief, Sir John French, took part in manoeuvres in Dorset, Wiltshire, Hampshire and western Sussex. Assembling the troops served as practice in mobilization of the Army in the event of war. Britain's first military airfield had recently opened at Lark Hill and aircraft were used unofficially for reconnaissance for the first time, but failing to impress. The exercises ended with a set-piece battle, which a German observer pronounced 'very pretty indeed. I would rather not say more'. Another foreign observer remarked: 'Your soldiers – oh, yes, they are admirable; but your generals is – pouf'.

Communications were still problematical, though cavalry on pedal cycles were much employed as dispatch riders and the sunny weather allowed heliographs to be used. About a hundred motorcyclists, some civilian, some military, were involved, acting as orderlies for the directing staff and scouting for both 'armies'. Wireless again failed to distinguish itself: in an exercise in August a detachment of the 18th Hussars had captured from the Red Force an unescorted Yeomanry wireless unit near Andover, together with papers revealing the basis on which the 'enemy' was operating. Later that month, the wireless units proved 'serviceable', one based 2 miles north-east of Salisbury contacting others at Semley (17 miles away) and Hamilton Camp (8 miles distant), though the Semley unit was unable to transmit back. *The Sphere* commented that 'wireless telegraphy was not altogether an unqualified success'.

Difficulties facing the soldiers in camp included roughs attacking them and hawkers selling them 'injurious trash'. (Permits were required to sell on War Office land and were not available to sellers of sherbet and lemonade – perhaps because the purity of the water they used was questionable.) Damage by such people was often blamed on the troops and some of the offenders were hunted down by bloodhounds. Two problems that had been noted as long ago as the 1898 manoeuvres recurred: the gaiters worn by some troops were too stiff and caused blisters, and some of the part-time soldiers' boots fitted badly, partly due to their having been worn by other people.

Writing in the *Wessex Divisional Journal*, Captain John Savile Judd of the 5th Duke of Cornwall's Light Infantry added to the reservations about Salisbury Plain: 'I have never noticed enthusiasm on the part of the Territorial soldier when it is announced that he is going to camp on Salisbury Plain.' He pointed out that the part-time soldier wanted to cut a dash in his uniform, but so commonplace were military men on the Plain that no one came running to see him. And many Territorials preferred camps within a reasonable distance of a seaside town, to which they could repair at the end of the day. (Braunton, Lulworth, Mudeford and Kingsbridge all hosted Territorial camps around this period, as did Blackpool, Clacton and Morecambe.)

In July 1911, 4,000 members of the Junior Division of the Officers' Training Corps (OTC) camped at Tidworth Park and Tidworth Pennings. The Senior Division was at Windmill Hill, near Ludgershall. For the Junior OTC, reveille was at 5.45am and breakfast at 7.30am, followed by field work from 9 until 1; after dinner there was a parade and further field work. There was as little ceremonial drill as possible. Postcard photographs of such camps

This card was posted on 4 August 1911, its sender stating that the damage is the result of manoeuvres on Salisbury Plain; 'you can hear the Guns all day', he wrote. The building was probably part of a farm near Market Lavington included in the War Office's purchase of land for training.

show fresh-faced and enthusiastic lads, many of whom would die as young officers in a few years' time.*

The major exercises planned for September 1911 were cancelled, ostensibly because of 'scarcity of water in Wiltshire', which meant (reasoned the War Office) that many acres of root crops could not be put at risk as they were of 'exceptional value' after drought had caused a scarcity of cattle food. (Rain fell on 13 September, two days after the cancellation was announced.) Certainly it had been a very dry summer, though in August the *Andover Advertiser* reported no scarcity of water on the Plain, thanks to the reservoirs serving the camping-grounds. Noting that the Army was limiting leave to just three days, the *Salisbury Times* was also sceptical about the reason for the cancellation. It was actually due to increased international tensions, with Germany having

---

*The Officers' Training Corps had been formed in 1908, its Senior Division being drawn from universities and the Inns of Court, the Junior Division from public schools. Before then, mass concentrations of, and extended camps for, individual cadet corps were rare, many schools running their own field days, sometimes with a few near-neighbours such as in July 1906 when Bradfield, Marlborough, Winchester and Wellington came together in Savernake Forest.

provocatively sent a warship to the Moroccan port of Agadir, a move seen by France as a threat to its interests there. As the latter's ally, Britain was poised to send an expeditionary force to northern France.

Furthermore, there was violent unrest at home in many industries, notably on the railways. This led to soldiers guarding railway installations and controlling riots elsewhere in the country, in some cases firing on crowds with resultant injuries and deaths. Workers on the Midland & South Western Junction Railway (MSWJR), which ran through Wiltshire, did not take part in the strikes, certainly in the Ludgershall area, from where special trains departed carrying troops to trouble spots. Bulford and Grateley stations on the London & South Western Railway also saw much military activity.

Though the Haldane reforms had led to a reduction of 37,000 troops in the size of the Regular Army, there was now a shortfall of some 8,000 men and, in an attempt to bring the Wiltshire Regiment up to strength, its 1st Battalion left Tidworth in mid-June on a 150-mile recruiting march lasting a week and extending to Devizes, Swindon, Chippenham, Trowbridge, Warminster and Salisbury.

August Bank Holiday in 1913 saw London Territorials arriving on the Plain for their summer camp, thirty-six trains leaving Waterloo, Clapham Junction and Nine Elms stations with 12,000 men. Among them was the London Regiment's new Hackney-based 10th Battalion (the original 10th, the Paddington Rifles, having been disbanded the year before), which camped at Perham Down. It comprised 800 men, with eight drill sergeants loaned from the Coldstream Guards for the occasion. The observations of one of its officers, Henry Prittie (later Baron Dunalley), were franker than most contemporary accounts, which, in the newspapers at least, tended to eulogize about the quality and bearing of Britain's soldiers. According to Prittie:

> What my crowd wanted was steady drill from dawn to dark every day, varied by a bit of musketry instruction. Instead of which – two days wasted by brigade schemes, one by a division day, another by night operation. Night operations for men who had never been out of sight of the street lamps of Hackney!

The battalion's guard-tent was usually packed with offenders, some of whom Prittie tried to punish with from ten to fourteen days' detention – until it was pointed out that certain of them were due to spend only eight days in camp and they would lose their civilian jobs if detained any longer. He halved the sentences and had no more trouble.

In 1914 a feature of the early summer camps was a larger proportion of Territorials staying for a full fifteen days, due to a bonus system introduced to discourage them from leaving after only a week. Growing friction between the European powers meant that a common topic in the camps through the summer was the likelihood of war; when the Wiltshire Yeomanry broke up in May

Few villages as small as Ludgershall had such a busy railway station. This extra-wide platform was designed for large numbers of soldiers arriving at and departing from their summer camps.

from its camp at Pyt House, Tisbury, few of its members doubted that the next time they would come together would be on mobilization.*

Mobilization

Late in July 1914, with international tension growing by the day following the assassination of Archduke Franz Ferdinand of Austria on 28 June, troops were preparing to concentrate on Salisbury Plain for the usual extensive summer manoeuvres. The British Government believed that cancellation of these would be seen as provocative by other nations, though it started to prepare to mobilize the Army on 27 July when Reserve officers were ordered to report to the British Expeditionary Force (BEF) headquarters. On the evening of Sunday, 2 August, all units training away from their usual bases were ordered to return. But already some 10,000 Territorials were assembling on the Plain for their annual camp, with the Home Counties Division assembling at

---

*The Royal Wiltshire Yeomanry was traditionally linked to Pyt House. During the Wiltshire farm workers' riots against machinery in the 1830s the most serious disturbance took place there, when the Hindon Troop engaged 500 rioters intent on smashing the machines that threatened their jobs. This action led to the Wiltshire Yeomanry becoming the first such unit to be awarded the title 'Royal'.

Bordon in Hampshire to march to Amesbury. So when Britain declared war on Germany on 4 August they were in a high state of readiness but often a long way from their home localities, so had to rush off to designated war stations to guard depots, stores, railway signal boxes and tunnels. Four Hampshire Territorial battalions left their camp at Bulford for Hilsea, near Portsmouth, but were back at Bulford within nine days (moving to Hamilton Camp on 31 August and to Bustard Camp a week later). The Army Service Corps' 12th Horse Transport Company from Portsmouth was at Tidworth, having a relatively easy life with officers and non-commissioned officers (NCOs) enjoying rambles towards the end of the day when they collected mushrooms and raspberries for their messes. It was ordered to return to Portsmouth as soon as possible, a trip of 55 miles that normally took two days by road, with an overnight stop at Romsey. The 12th completed the journey in one day, with only one horse going lame.

On 2 August the 14th London Regiment – the London Scottish – was experiencing its first night under canvas at Ludgershall Camp when, fifteen minutes after 'lights out', a bugler aroused the men to return to London.

John F. Lucy was based with the 2nd Royal Irish Rifles at Tidworth and, in *There's a Devil in the Drum*, described how he and his colleagues mobilized:

> We sold our own review uniforms to visiting contractors. We also sold our mufti clothing and boots, and the other few possessions we had. Property does not encumber Atkins [a popular term for the private soldier]. All superfluous Army peace-time gear was given back to the quartermasters; and ammunition, iron rations, jack-knives, and identity discs were received instead.
>
> We were all inoculated against typhoid, and some of the men fainted under the full shot. We were advised to make our wills in our soldiers' pocket-books, but the pay columns of these little books interested us more than the will forms. We would get higher pay in the field.

Also at Tidworth, Ben Clouting of the 4th Dragoon Guards was told to let everything go rusty and that nothing that reflected sunlight should be polished.

On 25 July the 4th Wiltshire Regiment had started its annual camp at Sling, the camping-ground next to Bulford barracks, and on 3 August was put under mobilization orders, expecting on the outbreak of war to go to its depot at Trowbridge. In fact, it marched to Salisbury, 'piled arms' on Cathedral Green, and entrained, without kit, for Devonport on 4 August, then travelled to Durrington (barely 4 miles from Sling) on the nights of the 9th and 10th; it then moved to West Down Camp on 16 September and on to Trowbridge on 15 October.

James McCudden was an engine mechanic at Netheravon Airfield where on 4 August his squadron mobilized for war and armed guards were placed in the aero sheds. Three days later he and his comrades were turned out to scour 'the country in the vicinity of the sheds to look for supposed spies who were reported to be prowling around with the intention of blowing up our sheds, but we did not find anything'. Later that month McCudden left for France, a journey that would lead to his being commissioned as a pilot, claiming fifty-seven victories over enemy aircraft and winning the Victoria Cross.

At the same time Reservists (men who had been discharged, in most cases in the past five years, and who comprised 60 per cent of the British Expeditionary Force preparing to leave for France) were reporting to their regimental depots and travelling on to join their units, often hundreds of miles away – all this at the busiest time of the year for the railways.

The London & South Western Railway provided thirty-eight special trains to convey the Home Counties Brigade of Territorials from Amesbury to its home stations between 10.20pm on 3 August and 3.30pm on 5 August – 14,000 officers and men, 1,387 horses, 310 tons of luggage, seventy-eight guns, 211 vehicles and 222 cycles were transported. During the fortnight of mobilization 632 special troop trains (not all carrying men from Salisbury Plain) ran over Great Western Railway lines, including 186 to return Territorials from their summer camps. Regular civilian services were maintained, only excursion trains being cancelled.

Perhaps officers and men of the 18th Hussars (Queen Mary's Own) had mixed feelings after they were told on 7 August that Queen Mary was coming to Tidworth the next day to bid them farewell before they departed for France. The gesture was well meant, but must have proved a distraction at a time when they were hastily mobilizing for war prior to landing in France on the 16th.

Quite how appropriate to war had been all the training of the past seventeen years on Salisbury Plain is debatable. There was nothing wrong with the basics: moving and deploying large concentrations of men; feeding them in the field; teaching them to shoot; and instilling *esprit de corps*. The mass movements of troops by train to and from manoeuvres had provided worthwhile experience, the rail companies coping well with mobilization, carrying 80,000 men of the BEF (and running eighty special trains into Southampton Docks on one day alone). This was facilitated by improvements for military reasons to Wiltshire's railways since 1897, which also proved invaluable through the war years.

The Plain had benefited artillerymen, who were to play a key role in the war, though the original land acquisitions had soon proved too limited for the increasing range of modern weapons. It had also enabled the comparative long-distance testing of communications, including the initially unreliable wireless. Fortunate too had been the locating there of airfields (at Lark Hill, Netheravon and Upavon), whose aircraft and pilots had eventually convinced a sceptical Army of the value of their cooperating with ground troops, even if, at the time that war broke out, little thought had been given to arming aeroplanes. Some attention had been paid to giving the different military arms a chance to work together and there had been a few experiments in billeting soldiers, albeit long after other modern European armies had been practising it regularly.

However, exercises on the Plain or other open spaces provided experience only in warfare on similar terrain. (Ironically, several soldiers were to compare ground close to the Somme with that of the Plain.) Luckily there was to be no test of the oft-voiced criticism that they never simulated preventing enemy troops landing, only fighting them in open country after they had done so. The commanding officer at Tidworth, Sir Henry Beauvoir de Lisle, may have spent three years from 1911 preparing his cavalry regiments for the war he saw coming, but after 1914 cavalry usually rode to battle and, being vulnerable to modern firepower, then fought on foot. For infantrymen, exercises on the Plain had consisted too much of moving around *en masse*, usually with little harassment from the opposing force, so as to be in position for an exercise or for their overnight camp, as had been planned several weeks before. Despite the role of trenches in the Crimean and American Civil Wars, there had been little practice in attacking defensive works or digging in. Artillerymen had been assessed on their gun drill and accuracy, but not so much on their ability to move their guns quickly into position or out of trouble. Machine guns had been little exploited, partly because their numbers were limited for budgetary reasons. Modern transport had not been tested in the field.

But then no nation had imagined a world conflict that would last for four years, feature static trench combat, introduce tanks and gas warfare, speed up the emancipation of women, depose several monarchs, lead to Germany and Russia becoming republics and kill ten million people.

For the next four years military training in Wiltshire would concentrate on preparing civilian recruits for the new type of war. First came the 'Kitchener' battalions, comprising civilians who had responded to the field marshal's famous appeal for 100,000 men, which was met so rapidly that there were acute shortages of uniforms, equipment and accommodation. Then in October 1914 arrived the First Canadian Contingent, hastily assembled from civilians and ex-soldiers, to be followed in mid-1916 by Australian and New Zealand troops, some with basic training back home but softened by the eight-week voyage, and others convalescing and retraining after wounds and sickness. The aim was to turn civilians into soldiers in only a few months, improving their fitness through marches, introducing them to basic drill and field craft, providing instruction in using firearms and the bayonet, and putting them through a few field exercises. The wounded and sick would have their progress assessed, the training increasing in severity as their condition improved and often irritating those who had already seen active service and thought they knew it all, or most of it. By the spring of 1915 the Army realized the significance of trenches in the new-style warfare and reflected this in training programmes, with most camps having intricate sets of practice trenches nearby where men spent a couple of days and nights to gain a foretaste of the real thing.

All this was in contrast to the carefully planned manoeuvres of peacetime and would take place throughout the year in all types of weather, rather than only in summertime. With its camping-grounds, improved railways and roads and proximity to the port of Southampton (and easily reached by two main rail lines from Plymouth, also much used for embarking and disembarking troops), Wiltshire may have seemed a suitable place to train the men of Britain's New Armies and troops from the Empire, but it was hardly to prove ideal.

2
The New Armies

Early Chaos

Many civilians who joined the Army early in the Great War found their induction and initial training disorganized, frustrating and demanding. This was particularly true of the very first recruits who with little or no hesitation gave up their jobs and enlisted, only to find that the Army was not ready for them. Many individual and regimental accounts of the war's early months cite examples of the chaos and discomfort of an organization used to handling 35,000 recruits each year being unable to cope with 500,000 in six weeks. By mid-September it had to look after almost a million men, compared with 125,000 before mobilization.

Small wonder then that many recruits not only had no huts, but also no uniforms and no equipment. Their very few serving officers and NCOs were augmented by those who had retired from the Army and who sometimes brought back outmoded forms of drill and tactics. Many new NCOs were selected on a provisional basis from the recruits themselves. Junior officers were mainly ex-public schoolboys, with experience of the Officers' Training Corps, but none of handling men from different backgrounds to themselves. At the OTC camp at Tidworth Pennings shortly after war was declared, Cyrus Greenslade, a lance-corporal in the Corps, was commissioned into three different battalions of the Devonshire Regiment. He was instructed to join the 10th Battalion being formed at Stockton in the Wylye Valley and on arrival was ordered to collect about 1,000 men, mostly in civilian clothing, from Codford Station.

A fine account of the trials and tribulations of a new soldier is that of Christopher Hughes, who lived in Wiltshire before the war, served with the Wiltshire Regiment and did his initial service in four different parts of the county. Aged thirty-four, he was an art master at Marlborough College who tried to enlist in the Wiltshire Regiment in September 1914, but initially had to settle for the Wiltshire Yeomanry. After some weeks (for most of which he had no uniform) in billets at Chippenham, where he received very little training, he was accepted by the regiment. After he had taken his fourth medical examination, consisting 'of a few vague questions as to how I was and as a test

The New Armies

Four newly raised battalions of the Cheshire Regiment arrived at Codford in late 1914, including the 13th – the 'Wirrals' – shown here in makeshift uniforms. Unlike some units they have rifles, though these are Lee-Metfords, which the Regular Army had phased out almost twenty years before.

for the eye sight being asked how much I could read of the cover of the Army list held at arms' length', he found himself back in Marlborough with its new 7th Battalion. Its colonel was Walter Rocke, who had retired from the Wiltshire Regiment nine years before to become Honorary Colonel of the Leicestershire Militia. Shortly after the start of the war he had presented himself at the 5th Battalion's base in Tidworth in full uniform and was given the command of the 7th.*

The 7th Wiltshire's officers had the typical backgrounds of a Kitchener battalion:

> The Colonel and the Majors had retired from the army some years. Of the four Company Commanders one was, at the beginning of the War, Sergeant Major of the Depot, one a retired Sergeant Major, and two were reported to have had some training in the Colonies but we Subalterns regarded this as somewhat doubtful. The Adjutant came from the first Battalion.
>
> Of the Platoon Commanders some had been in their School OTCs and the others had no training whatsoever ... The Battalion Sergeant Major was a marine, the Company Sergeant Majors all men who had retired some years

---

*Infantry regiments added new battalions – such as the 7th Wiltshire – to their peacetime strengths, and Territorial battalions adopted a new form of numbering: the original battalion would become, say, the 1/1st and a second would be raised – the 2/1st – and then a third and sometimes a fourth.

The New Armies

as Sergeants. The Company Quartermaster Sergeants were taken from the ranks, Bank Clerks, and Clerks of other kinds. Except for a few old soldiers who had retired as Sergeants and Corporals and had rejoined, the NCOs were all drawn from the ranks and had enlisted at the same time as the rest of the men...

So the only serving soldiers were the Adjutant and one Company Commander. And there were some fifteen hundred men absolutely new to military discipline.

One instance of this unfamiliarity with Army life was when a recruit called out, during the company standing at attention, 'I say mister, I've got such a pain in my belly may I go 'ome.' (At Tidworth another unit of recruits on a route march is said to have broken ranks as it passed a pub, disappearing inside under the eyes of its startled and angry NCOs.)

The first aim of Colonel Rocke was to get the men fit, with two physical training parades every day (which included all battalion officers except company commanders) and route marches, building up to a daily average of twenty miles, except on Sundays. On most fine, and some wet, days, there was field training, but though the tactical movements provided 'quite a good drill', they belonged to the period before the Boer War:

> We would assemble on a hill some 2 miles from our objective; the Company commanders would receive their instructions, and usually had very little time to convey them to the Platoon commanders who had none at all to instruct their platoons...
>
> We would start our attacks in long extended lines with the men some ten paces apart, and therefore very hard to control, no hedge or fence was allowed to destroy our straight line, all open ground and all skylines had to be crossed at the double. When the Colonel was in the bottom of a valley everything seemed skyline to him, so we had to run everywhere, it may be imagined that our attacks were carried out at a tremendous pace.
>
> The first extended line formed the firing line and this was built up during the advance by the supports and reserves joining it, then a short distance from the objective bayonets were fixed and the Officers drew their swords and with loud shouts dashed into the position. The attack was always frontal, sometimes officers attempted to lead platoons by covered ways on to the flank of the position to be attacked but these efforts were not received with favour.
>
> Sometimes men carried poles to which were attached strips of linen to represent the enemy; on one occasion a short-sighted officer with the leading platoon took the battalion into a spirited attack on washing hanging in a garden. Miles of trenches were dug, all on reverse slopes because contemporary wisdom had it that no visible trench could be held under shell-fire.

Ten years later, Hughes observed that 1915 instruction related to set battles in which the only man to really matter was the senior commander:

> Modern weapons have pushed him into the background and thrown a much greater responsibility on the junior commander and the NCOs. It was in the failure to instruct these that was the most serious fault of our training.

When it was wet, officers had to give talks to their company or platoon:

> One officer lectured on knotting and got so confused that a man in his platoon who had been a Boy Scout had to come up and finish the lecture for him.
>
> Another, after a late night, recalling a lecture imperfectly heard some time before, spoke on march discipline. He told them [his men] that when on the march if thirsty to put pennies in their mouths (he should have said pebbles) and several suggestions of a like nature; his company commander had to lecture the next day to correct the errors.

It was discovered that no man in the battalion knew anything about scouting, so the equally ignorant Hughes was told to instruct on the subject. He was similarly deputed to teach signalling and after lecturing for some time discovered an expert telegraphist able to send Morse at some twenty-five words a minute, contrasting with Hughes' rate of 'a word of three letters a minute, with luck'.

There were the long hours of boredom and routine, abruptly interrupted:

> early morning parades, long marches to position, long hours of waiting, sudden orders, no time to explain or be explained to, confusion and more confusion, and marching home in the evening, and pow-wows with blame for everybody, the only hope was to keep out of the way of those in authority, the most dangerous enemy was in front, we took cover far more from the Colonel than from the enemy.

In April 1915 the 7th Wiltshire marched to the regimental depot at Devizes, where it spent ten days before moving to Sutton Veny. It had raised a good band that was sent out with a platoon to seek recruits in the towns and villages. Hughes led one recruiting march from Devizes to Bradford-on-Avon, which brought in fourteen volunteers. Among them were 'two village idiots about 4 feet high who wanted to join the Life Guards'. A second march, from Sutton Veny to Salisbury, brought in thirty-one, 'who were a better lot', though most eligible men concealed themselves when the recruiters approached.

'Our best work was done on Market Day after the public houses closed for curiously enough, as men became more inebriated they became more patriotic,' noted Hughes. Even after men had agreed to join the Colours there were problems as they were marched off to the station to take the train to the regimental depot in Devizes; parents, wives and sweethearts would try to snatch the recruits away.

The New Armies

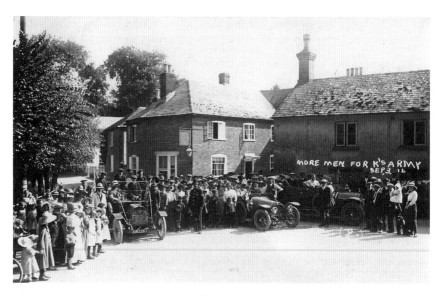

Local worthies in their cars encourage local men to enlist outside the Catherine Wheel Inn in Shrewton in September 1915. The photographer's crude caption reads 'More men for K's [Kitchener's] Army', though how many signed up is not known. The Darracq car on the right belongs to John Hayward of Maddington Manor, Shrewton.

Recruiting meetings were held in Salisbury market place and were sometimes addressed by the Bishop of Salisbury. But, thought Hughes:

> What a waste of time was all this recruiting and meetings. The meetings were usually attended by women and old men and addressed by more old men who by heroic phrases and tales of the glory of war sought to gain recruits. I think everyone not of military age was anxious that all who could should serve.

The 7th Wiltshire finally got modern rifles in August 1915, eleven months after being formed and only a few weeks before it left for the Western Front on 22 September, eventually serving in Salonika. Hughes ended the war a major with a Military Cross.

By late 1915, the enthusiasm and patriotism that had been the hallmarks of the Kitchener soldier had faded and, as more men were needed and were not forthcoming, conscription became inevitable. Parliament agreed to it for unmarried men in January 1916 and for married men in May. Local newspaper reports of tribunals hearing appeals against compulsory enlistment suggest many people had to be winkled out of civilian jobs.

The method in which units of recruits were organized also changed and local allegiances all but disappeared, though from the start many 'county' battalions had contained only a proportion of local men. At first they were inducted at the regimental depot and then, as part of a 'service battalion', moved to an area such as Salisbury Plain for, typically, three months' preparation for active service. The battalion would then supply drafts of men to fighting units. But this system was unable to cope with the number of conscripted recruits, so the 'Training Reserve' of 112 numbered battalions was formed to provide reinforcements for service battalions serving overseas. Later on, some of these Training Reserve units became 'young soldier' and 'graduated' battalions.

The Times struck a curious note when, after visiting Salisbury Plain, a correspondent reported on 25 May 1916 on the type of recruit who was now being transformed there in weeks from a sophisticated civilian into a trained soldier ready for active service:

> A newly promoted bombardier approaches his subaltern and saluting, asks if he might be allowed to bring his car into camp. Near a large training centre in the Southern Command there is an excellent golf course and every evening one may see two limber gunners, a driver and a battery telephonist play a friendly foursome. At least one of the four men has played against Braid and Hilton [celebrated golfers of the period]. Coming back to camp the foursome is joined by two other men in gunners' uniform, and talk turns to the preservation of game. The Number 1 Gunner of Number 3 gun tells the Number 3 of Number 1 gun just where and when he liberated his last lot of salmon fly in his own stretch of fishing in the North.

The report did not confirm that the men were conscripts but hinted that they could have enlisted a little earlier. It conceded that they could cause problems for artillery instructors when they asked if it would not be easier to work out the battery angle by using a subversed sine: 'the questioner may have been a qualified surveyor or honours man in mathematics'.

By mid-1916, most of southern Wiltshire was one vast training ground, dotted with hutted camps. The new hutments housed at least 120,000 men, with thousands more in the pre-war barracks, in billets and under canvas. The county had become the training ground of the British Empire, containing a tenth of all troops based in the United Kingdom. Many were straight from civilian life in Britain; some, from overseas and having done their basic training at home, were being prepared for the Front Line; others had returned from there to Wiltshire to learn new skills or to recover from sickness or wounds. The instructors now included men who had tasted modern warfare and who knew how inappropriate the pre-war manuals were (though the battle-hardened would always argue that no training programme ever related to the real thing).

The regime at Sand Hill Camp near Warminster in early 1916 was typical. Reveille was at 5.30am. As the men were dressing they were given a cup of lukewarm coffee known as 'Gunfire' and then had a 4-mile run, followed by a very modest breakfast. Then, on the parade ground, would follow 'Swedish drill' (physical training), marching and handling rifles. There were route marches and exercises, the latter involving mock attacks on often non-existent enemies. All this had been done before the war, but now the men were trained in trench life, even being given a whiff of gas. From late 1916 gas rooms or chambers were established at Bulford, Chisledon, Durrington and Fovant camps and at Tidworth Barracks, through which soldiers wearing masks would pass so as to accustom them to what they might experience in the trenches in France. The New Zealander W. J. McKeon recalled:

> We were introduced to lachrimatory [*sic*] or tear gas and had to face this with and without masks, just to get the feel of it! One particularly foul trick which we had to suffer was entering a sloping trench full of tear gas, running through it and up the other slope to ground level again. Completely blinded by the stinging gas, we staggered down the trench in an effort to get out as quickly as possible. In the lowest part of the trench some fiend had half buried a two-gallon petrol can, over which we tripped and sprawled into the bottom of the trench. Gasping and smarting, we staggered on, emerging at the other end done to a turn, to face the unfeeling laughter of the instructors who stood clear of the stinging atmosphere.

Rifles, Bayonets and Bombs

The pre-war British Army prided itself on its rapid rifle fire, with an average soldier being able to sustain aimed fire of fifteen rounds a minute and fire a 'burst' of up to thirty-five in the same time, a skill that in the first months of the war is popularly said to have made the Germans wonder if machine guns were being fired at them. But the New Armies' preparation suffered from a lack of rifles, the standard model in 1914 being the Short Lee-Enfield. For much of the war many recruits were not issued with it until late in their training or even not until they got to France, making do with the older Long Lee-Enfield, the Canadian Ross Rifle or Japanese Arisaka. Some men early in their training had to use Lee-Metfords, phased out from the Regular Army from 1895, or even wooden dummies. Whereas the pre-war soldier would have fired 250 rounds in training, the figure for some wartime recruits could be as low as a tenth of this, and inevitably standards slipped.

The Salisbury Plain rifle ranges were fully used, sometimes having to accommodate men from camps near Winchester as well as those based locally. There soldiers would fire at distances of 100–600yd at targets in a trench deep enough to accommodate two men *in front* of each target. (Lieutenant Ross Briscoe of the First Canadian Contingent was accidentally shot during rifle practice at Sling Camp on 6 January 1915 while standing in a pit *behind* the targets, where he was at risk from deflected bullets; he had been warned about this in December.) One man at the pits would use a disc on a long pole to signal the accuracy of each shot and a red flag to signify a miss; the other would paste over the hole made by it in the target.

Every officer and man had to train with the bayonet so that 'great havoc was committed upon a row of suspended sacks, usually decorated with ferocious representations of Hun physiognomy or facetious legends'. The Army believed that use of the bayonet instilled offensive spirit; certainly the training enabled the men to shout insults and violently lunge and twist the bayonet, whereas for most of the rest of the time they had to exercise self-control. Much time was devoted to familiarizing men in the bayonet's use – too much to some people's way of thinking, considering that in actual fighting few men were wounded or killed by it, partly because there were not many opportunities for close combat. But then the training was easy and cheap compared with expending ammunition on ranges that were much in demand.

If the bayonet was less used than anticipated by the training manuals, the 'bomb' or grenade soon had an important role, both in attack and defence, in the new style of warfare. In 1914 the British Army had only a stick grenade, whose 22in length made it a liability to throw from the confines of a trench. The first troops in France fashioned their own grenades and the 2nd Hull Service Battalion was taught how to make homemade bombs from jam-jars and condensed milk tins when at Hurdcott Camp in late 1915; they were judged as lethal to their makers as to the enemy and there were several fatalities. That year the Mills Bomb was introduced. The best of a variety of designs developed in the war, it was a cast-iron 'pineapple' with a striker held by a lever (in comforting contrast to models that required a fuse to be lit – no easy matter in wet conditions) and secured with a pin. A competent man could manage a 100ft throw, but had to take cover immediately. Most camps had sections of trenches for training in throwing, first of all dummy bombs and then the real thing. In no part of the training was self-control more important than when on the 'bomb' course. Defective bombs and inexperience resulted in more accidents than with any other form of activity.

During bomb-throwing practice at Tidworth in October 1915, a man lit a fuse and discarded the match into 15lb of powder. Three men of the 23rd Royal Fusiliers had their eyes damaged and others suffered discoloured faces; two specialists from Harley Street came to Tidworth to attend to the former. The 23rd's bombs were made from iron cylinders filled with gunpowder, each end being plugged with clay and fitted with a timed fuse.

On 7 November 1916 Lieutenant Cyril Carey of New Zealand was killed at Sling Camp when a man failed to release the grenade correctly and it lodged on the inside of the bombing-bay. Carey forced the man down to the ground and attempted to throw the bomb away but it burst within 2ft of him, giving him severe head wounds from which he died ten hours later. That same month, Corporal Arthur Button of the 51st Battalion was killed and ten other Australians injured at Codford Camp. A private under instruction had held on to the bomb for too long and thrown it incorrectly so that it fell into the next bay, where ten men were waiting to throw. It exploded prematurely, in three seconds rather than the usual four and a half. The officer in charge of the practice told the coroner's inquest that such incidents happened 'fairly often' – though hopefully with lesser consequences.

The job of instructor was a dangerous one, as the Australian Thomas Kermode of the 8th Training Battalion and an instructor at Hurdcott, explained in January 1917:

> I have been acting as a bombing instructor in bombing bays. We take say 60 men and from behind breast works they throw, under our direction, the famous No. 5 or Mills grenades. Before throwing we learn them how to assemble the bomb, then to throw them. Now and again what we call a 'dud' occurs, that is it is thrown with the safety pin out, the cap strikes, but the explosion fails. Each instructor has therefore to go out from cover and pick this 'dud' up. We are not sure whether it will go off before we get to it, but a pose of courage and careless indifference is shown, or what would the men think. I had my first experience of doing this today. I can tell you the blood tingled in the roots of my hair, for I knew if it did go off as I approached it, my face would get the lot. Still, I showed no hesitation and it was alright. But it is no mugs game.

There was an unusual fatality at Sutton Veny Camp in 1916 when George Pearce of the 2/18th London Regiment (the 'London Irish') was operating a West spring gun, comparable to a Roman catapult and used for throwing bombs. Pearce, thinking he had forgotten to light the fuse on a dummy bomb, bent over the gun; as he did so, it went off and the bomb and cap shot up and hit him in the face.

In October 1915 Private William Shannon of the 15th Royal Scots was killed attending a demonstration at Sutton Veny on how to use explosives to destroy barbed wire. Fifteen ounces of wet gun cotton and one ounce of dry gun cotton were attached to about 16in of iron railing, all of which was tied to a barbed-wire entanglement. It appeared that no direct order to lie down was given to the audience, who were merely told that 'men who wish to lie down can do so'. Shannon, a trained grenadier, had stood up before the explosion.

Trenches

Practice trenches were being cut on Salisbury Plain by at least 1902, when guns and howitzers were fired at three S-shaped trenches 4ft deep. But the

The West spring gun comprised a bank of springs, a cocking lever and a throwing arm and could propel a bomb up to 250yd. Some soldiers thought it was more dangerous to them than to the enemy and its use was officially discontinued in July 1916. This photograph shows men of the West Yorkshire Regiment demonstrating it at Fovant Camp.

indiscriminate digging of trenches was expressly forbidden in the War Office's pre-war standing orders for the Plain. In 1909 (when that year's Field Service Regulations noted that trenches were an essential ingredient of defence), digging was allowed in certain specified areas. Between Tidworth Cemetery and the River Bourne 500 acres were put aside for entrenching practice and Haxton Field Barn (south of Everleigh) was placed 'in a state of defence as an entrenched locality'. Temporary entrenchments were allowed to be made on manoeuvres, but to avoid heavy compensation claims by landowners they were not permitted where a chalk subsoil existed close to the surface, because, when they were refilled, the chalk would become mixed with the top soil, greatly reducing its agricultural value. This restriction precluded most of the Plain.

Some members of the First Canadian Contingent lamented in 1914–15 that they were not allowed to dig trenches lest horses of the local hunt fell into them. In fact, the Contingent's diaries do mention the digging of trenches, which were then quickly filled in. That winter was so wet that permanent excavations would have flooded, as other troops found for themselves a year or two later; and in some localities solid chalk bedrock was too close to the surface to allow excavations of sufficient depth. Evelyn Southwell, writing from Windmill Camp in the dry summer of 1915, described digging:

> in a valley one mile off with one inch exactly of brook in the middle of it . . . we went and dug trenches in it, which was rather a desecration and frightfully foolish, for of course we struck water two feet down, and the digging was much harder than even our ordinary chalk digging, and that is no joke.

The 7th Wiltshire Regiment was billeted in Marlborough in late 1914 and its men dug trenches about 1,500yd north-west of Rockley, 'such as are being used at the Front and are worth inspecting from an instructive point of view', reported the *Marlborough Times*. Field Marshal Lord Methuen visited them on 23 December. However, a contemporary postcard photograph shows the design to have been very simplistic and of a type criticized in the 1915 staff manual *Field Entrenchments: Spadework for Riflemen*, in which a drawing showed how *not* to construct a trench. As with the Rockley example, there were too many men crowded in a straight line who could be wiped out by enfilading enemy artillery fire (that is, from one end of the trench to the other); it was too narrow to allow communication and movement; there were no support, communication or rear trenches; and the defenders would have faced downhill 'attacks'. (Today in summer on the site there is a profusion of poppies, poignant symbols of the spilled blood of the men who trained there.) A similar trench – one narrow, featureless excavation – was dug at Fovant in autumn 1915 and served only to give the men practice in jumping out of it with rifles.

But the 7th obviously remembered its time at Rockley when serving close to Lake Doiran in Macedonia as its members named a hill after the place. Similarly, they named other features Marlborough Hill, Swindon Hill and Wylye Ravine.

Soon the demands of war and the realization that much of it in France would be static and fought in trenches meant that every major infantry camp on the Plain had an increasingly complex set of trenches nearby, in which units might spend several days and nights to gain experience of what living in them would be like. Rations would be brought up from the rear and emphasis placed on units relieving each other at night. A typical night exercise at Sand Hill in 1916 was based on a relatively modest trench 50yd long and 5ft deep with no dugouts or shelters. One company would defend the trench against the assault of another; with no means other than blank ammunition to repel the attackers there would be a great deal of hand-to-hand scrapping in the trench itself. The Australian 38th Battalion 'lived' for several days in a trench system near the Bustard Inn. As Eric Fairey records in the 38th's *Official History*:

Men of the 7th Wiltshire Regiment in a very basic trench near Rockley. It is facing uphill, the belief in 1914 being that trenches had to be out of sight of the enemy so as not to attract shelling. The road from Marlborough to Wootton Bassett can be seen in the background.

> Rain swept the open country and poured into the white-chalk trenches. When at night several companies entered the trenches to take up their positions, men floundered through pools of whitewash and got covered with sticky white mud. Verey lights went hissing up through the driving rain, to illuminate a dreary landscape. Rifles cracked, and the dull detonations of hand grenades momentarily drowned the angry hissing of the rain.

Significant trench systems of the Great War in Wiltshire included those at or near (with map references shown):

* Beacon Hill, east of Bulford — 215442
* Bedlam Buildings (site), south-east of Tidworth — 252474
* Bustard — 080473
* Bustard (Shrewton Folly) — 087468
* Chapperton Down, west of Tilshead — 990475 & 994480
* Imber — 970465
* Knook Down — 965455
* Market Lavington — 043518 & 048518

* Slay Down, north-east of Tilshead   084511
* Whitefield Hill, south-east of Chisledon Camp   211768
* Yarnbury Castle   041406

The Knook Down and Chapperton Down systems appear to have been deliberately shelled to demonstrate the effects of artillery bombardment. A century later, erosion and silting have reduced earthworks to at best bumps and scrapes, to the extent that visits to the above sites can disappoint. The most obvious traces (said to be among the best in Britain) are on Beacon Hill, between Park House and Sling camps, where the original system included island traverses, communication trenches and a reserve line that made use of a Bronze Age linear ditch. The 'island traverse' was an elongated earth or sandbag buttress between two sections of trench, the aim being to reduce the effect of shellfire. The 'Bedlam practice trenches' were not ironically nicknamed for the confusion that might have afflicted soldiers training in them, but were called after Bedlam Buildings (a farm or barn) east of Tidworth Park. On Whitefield Hill a large crater is still visible, the result of a mine being set off, presumably to lend realism to training; ploughing has removed all traces of the nearby trenches.

During the war the Army thought that chalk cob might be used for the rapid construction of trenches in France. It had long been employed in many Wiltshire villages to build houses and erect walls and consisted of chalk and water mixed into a paste then slapped into position; a course about 18in high could be added at a time, any more causing the paste to start collapsing under its own weight. The mixture took a couple of days to dry and afterwards had to be kept dry at bottom and top – all conditions that would seem to preclude it as suitable for trenches. Nevertheless, the Army rounded up local practitioners to introduce soldiers to the technique, but the cob proved too fragile and prone to disintegration.

Royal Inspections

A few days before a division left Wiltshire for active service, it was customarily reviewed by the King, an occasion viewed with mixed feelings by the participants. Many were relieved that their period of training was over, though doubtless some were not looking forward to what lay ahead. As with inspections part-way through the training by senior officers and visiting dignitaries (such as, in the case of Dominion troops, politicians from home), there might have been a time-consuming rehearsal, though sometimes troops marched some distance to the review ground, perhaps camping in the open overnight, and were marshalled into position on the day itself. Drum-Major Jack Paterson wrote to his mother about the royal review at Bulford on 17 April 1917, his letter being published – with camp names omitted for security reasons – in Australia's *Western Argus* of 11 September:

> Just a line to tell you of our march of seventeen miles to be reviewed by the King. We were up at 5.30 on Monday morning, it being not a bad morning for a start. At 7 a.m. we marched out of camp, our Battalion Brass Band, with myself at the head, leading the battalion . . . At 11.30 we halted for dinner, just a mile or so this side of —— camp, and after dinner we marched into —— . By this time the rain was pouring down something terrible. We were camped at —— for the rest of the day and night. It was an awful night, cold as it knows how to be over here, and the wind was whistling. Reveille was blown at 5 a.m. next day, and we were marched out of camp over on to the review ground, about five miles distant, at 8 o'clock. It was a marvellous sight to see the long columns of troops coming from all quarters, up hills and down dales . . . There were over 500 bandsmen in mass; and all the side drummers and drums in front of them, and the drum majors in front of them again. We stood for one hour and ten minutes, in cold, bitter wind; at attention while the whole parade passed us, but we had the consolation of knowing that King George had to sit to attention on his lovely big black horse and take the salute too.

After such events it took a while for the troops to disperse, with those from some distance away starting the march back that same day. All this was days before the division left for overseas, with checks having to be made on the men's health and equipment and plans drawn up for them to march from their camps to catch a series of trains, not all of which left from the nearest station. And sometimes in the last week or two there was a change of mind about where they would be going, with several instances of tropical kit having to be exchanged for that designed for Europe.

The King's military visits were not made public in advance, but inevitably news leaked out and on 23 June 1915 a crowd gathered to see him arrive at Ludgershall Station for a three-day visit. At 12.50 the royal train backed into the long station bay as the King lunched. At nearly 2pm, he left on his black charger, with local National Reservists forming a guard of honour down Station Road.

'The battalions who have now been here for some months hope that this visit of His Majesty is the preliminary to being moved elsewhere, as after a time the most beautiful scenery begins to pall,' noted the *Andover Advertiser*. That day the King inspected the 19th Division. On the 24th, still at Tidworth, he inspected the 15th Division, and then at Knighton Down, Lark Hill, the 20th (Light) Division, before travelling to Normanton Down, south of Stonehenge, where the 18th (Eastern) Division was drawn up. On the 25th he inspected the 26th Division; then he returned to Tidworth for the 37th

The New Armies

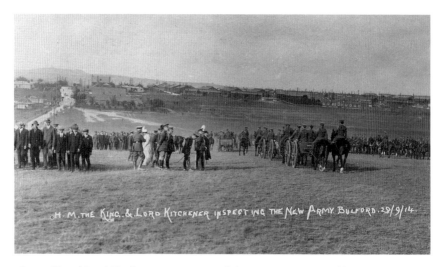

George V and Lord Kitchener inspect men of the New Armies at Bulford on 28 September 1914. (The card's caption gives the wrong date.) Some have their new uniforms, but others, on the left, are still in their civilian clothes.

Division. Later the King sent the usual complimentary messages, that to the 18th Division stating: 'Your prompt patriotic answer to the Nation's Call to Arms will never be forgotten. The keen exertions of all ranks during the period of training have brought you to a state of efficiency not unworthy of my Regular Army.'

One wonders if these plaudits were sufficient reward for the 11th Argyll & Sutherland Highlanders, who had left Chiseldon on 23 June and marched 11 miles to a bivouac camp at Wilcot, a mile north-west of Pewsey, then the same distance the next day to Sidbury Hill, Tidworth, where, as part of the 15th Division, they were reviewed by the King. That same day they marched back to Wilcot, on the 25th completing their return to Chiseldon, where they prepared for departure to France. They may have been ruefully aware that Chiseldon and Ludgershall, just 2 miles from Tidworth, had a direct rail connection, but no doubt the marching and overnight camps were regarded as useful training.

Even when close to the review area, troops could face long waits: on 27 September 1916, despite there being five approach roads to the review ground at Bulford, the men started from their nearby camps at 7.15am, four hours before the King was due to inspect them. Australian and New Zealand troops, 26,700 in all, formed a line 2,200yd long and took two hours to march past to the music of sixteen massed bands. However, many of the participants were

The King and Queen enter Trowbridge Town Hall during their visit to Wiltshire factories on 9 November 1917. Note the wounded soldiers in their wheelchairs.

not happy; the Australian commander, Major-General John Monash (later acclaimed by some as the ablest of all wartime generals), complained that 'we are being inspected to death and it does disturb the training so'.

For his visit of 12 February 1917, the King alighted at Dinton Station and motored to Fovant Camp to inspect troops and present medals before being driven to Wilton. 'The route was efficiently kept by the civil and military police, who diverted or stopped all traffic,' reported the *Salisbury Journal*. 'The procession of staff cars, as it swept through the villages, was cheered by the inhabitants and greeted with flags waved by school-children assembled at points of vantage.'

The best publicized visit to Wiltshire was on November 1917, when George V toured factories contributing to the war effort. Because of the non-military nature of the visit, local newspapers were able to give details in advance and publish full reports afterwards.

Lesser personages conducting reviews were not always well regarded. When Field Marshal Lord French visited on 22 February 1916, A. F. Barnes in *The Story of the 2/5th Battalion Gloucester Regiment* noted:

> At 1.20 we marched out of the huts and were drawn up on a hill side; snow was falling and a perishing gale blew: we stood for two hours frozen to the marrow. Sir John French came around at 3.20 p.m. and had a look at us. He

was a small man with a white moustache and a red face: he wore a big fur coat which hid most of him.

In May 1917 Victor Smith of the 66th Battalion, Australian Imperial Force, noted: 'We went across to Tidworth to be inspected and reviewed by some American merchant and he was cheeky enough to review us in Civvie clothes. Rotter. We were dogged tired when we got home.'

The march to Tidworth would only have been one of a couple of miles, from Smith's camp at Perham Down, though the return journey would have been uphill. At that time Smith was graded 'B1A2' – reckoned to be fit enough to go to an overseas training camp in three to four weeks – but he was finding the regime rigorous, several times recording in his diary how tired he was. His condition was not helped by having to sleep in a tent. The month before, he and his comrades had been 'issued with full equipment and were also inspected by some Red Tape Generals. One of them looked like a bag of chaff, tied up in the centre with a rope. He was as good as a circus.'

When the Duke of Connaught (third son of Queen Victoria and the last member of Britain's royal family to be a career soldier) inspected ANZAC troops on 20 August 1917, Ira Robinson of the New Zealand Rifle Brigade admitted beforehand (as recalled in his memoirs, *Dear Lizzie*):

> I am living in hopes that I will not be in it, as you can see more if you are a spectator. Anyway, being in it yourself is no good as you have to stand like a stuffed monkey while a dozen or so of the heads, or tin hats as we call them, glare at you and try and find some fault with the way your boots are laced up or something else equally as trifling.

NZ Truth reported on the Duke's visit in its edition of 1 December 1917:

> When Royalty visits Sling Camp to review the New Zealand troops there are always great preparations in the way of extra drills and parades. On the occasion of the King's visit about a week was spent on rehearsal; one of the days being Sunday, divine devotion was cut out. For a review recently held by the Duke of Connaught, preparations were not so extensive, but very amusing to the men all the same. They were paraded in all sorts of dress and equipment. First of all they paraded with the valise, which carries overcoat, underwear, shirts, soxs, etc. About an hour after they had to take these off after having adjusted them correctly and parade without them. Later on they paraded with them on again, when it was discovered that the entrenching tools were not right. So these were altered. Finally, it was decided to put the entrenching tools back as at first, and the men – 10,000 of them – were moved off to the ground for inspection by the Duke. They lined up there at two o'clock and were kept waiting until three, when H.R.H. arrived in a motor car. He had a look at the front ranks and did not bother with the rest. Then the troops marched past,

and H.R.H., later on, expressed his satisfaction through group orders, stating he was sure that such a fine stamp of men would do credit to the fair name of New Zealand.

Departing for War

Units often had only a week or two's notice of their departure for active service, though they would have known that their training on the Plain was likely to last about three months. Inspections of the type described above would have confirmed that a move was imminent, although the destination was known only to senior officers (albeit the subject of much rumour and speculation among the rank and file) and was subject to change at the last minute. One chore was for the men to sew patch pockets on the inside of the tunic, one for a field dressing, another for an anti-gas helmet.

Usually, although not always, a few days' leave were granted, spent by many British soldiers at home – and travelling there and back, which caused resentment among those with long journeys. Men of the 17th Durham Light Infantry at Fovant had hoped for forty-eight hours' leave, but their passes were cancelled at three hours' notice and they were ordered to Egypt – instead of France, where the 17th's advance party was already heading. Troops from overseas opted for London where, many memoirs relate, it was not a question of seeking female company, but rather one of fending it off. When back at camp the men packed all spare and unwanted gear and personal belongings into kitbags and were given receipts. On his return to New Zealand, W. J. McKeon was notified that he could collect his from a specified depot and was impressed to find it there, with its contents intact.

The outgoing soldiers were meant to leave their huts in a reasonable state, but this was not always done. The 2/15th London Regiment and the 14th Royal Warwickshire Regiment arrived at Sand Hill and Codford respectively to find their huts in a filthy condition. When it was the 14th's time to leave, they had the usual problem of disposing of personal and battalion possessions. The barrack warden (usually a former warrant officer or senior NCO who was responsible for cleanliness and equipment in the huts) tried to confiscate private property, much of which had been donated by the citizens of Birmingham. Eventually most of it was sold to Salisbury tradesmen, but there was a long argument with him over the battalion flag. When the First Canadian Contingent left Wiltshire in February 1915 its members saw local tradesmen descend on their huts to help themselves to what had been left behind. Lieutenant Charles Hennessey had hired a gramophone and twelve records from a Warminster shop, leaving a £5 bond as surety. He had time only to put a note on the items asking the next occupants of the hut to return them.

Often preceded by a band, the soldiers marched to a railway station – not always the nearest to their camp – to be taken to a port for embarkation overseas. Ken Murray of the 38th Battalion, Australian Imperial Force, recalled departures from Fovant in early 1918:

> It's an even more impressive sight to see a small draft go out than to see a troopship depart. The C.O. always makes them a touching little speech and the boys march up to the rail head with a yelling crowd of cobbers darting round them like flies buzzing from one chap to the other giving advice etc. and generally making pests of themselves. And always standing at a discreet distance is a young mob of girls and women with whom the chaps have made themselves acquainted during their stay. After all the goodbyes have been said, all the advice given, and the cheering over, the fussy little light engine grinds its way slowly around the curve.

Though very few British soldiers appear not to have returned to camp prior to their unit moving overseas, inevitably there were some late arrivals and these would be taken on to the port of departure by a junior officer who had been left behind to complete the procedures of his unit departing. The 14th York & Lancaster Regiment at Hurdcott Camp was given leave from 22 to 25 December 1915. It was to entrain at Dinton on the night of the 26th, by when ninety-two men had failed to return. Two officers met twenty-eight of the latecomers at Salisbury and took them by taxi to Dinton. A second lieutenant was left behind at Hurdcott to bring on other absentees by another ship, forty-five men making it to Devonport and three more arriving in the charge of another second lieutenant. (Then the ship was unable to sail because of a shortage of stokers and the battalion was asked to provide volunteers!)

3
Building the Wartime Camps

The Rush to Construct

Before the Great War, barracks in the United Kingdom provided quarters for 175,000 soldiers. On 6 August 1914, Lord Kitchener, the new Secretary of State for War, called for 100,000 men aged between nineteen and thirty to serve in the Army for three years or the duration of the war – the first of four Kitchener Armies to be raised in this way. Within a fortnight, the 100,000 target was met and by mid-September half a million men had enlisted (a figure doubled by the end of November). These men needed to be accommodated, fed and trained.

As a first step, soldiers' families were moved out of married quarters, increasing barracks' capacity to 262,000. Tented camps were erected and schools and empty factories were taken over as temporary accommodation. On 12 August the Directorate of Fortifications and Works was asked to submit plans for a standard hutted camp that could house one battalion at a war strength of 1,000 men. Within two days (having already done some preliminary work), the Directorate had produced plans embracing seventeen different designs, including those for officers' and sergeants' messes, a recreation hut and a large central cookhouse with a dining-hall for a half-battalion on either side. The staple unit was a hut, 60ft by 20ft and with an average height of 10ft, providing sleeping quarters for twenty-four men and an NCO. (Significantly, there are many references to upwards of thirty men to a hut, suggesting individuals' space had been compressed.) Forty such huts catered for a battalion's rank and file and could be built for an estimated £15,000. By 17 August the proposals had been approved by the Army Council and a construction programme set under way.

In Wiltshire, all the established camping-grounds were immediately brought into use, with British recruits at Park House and Perham Down, New Zealanders resident in this country starting to build huts at Sling Plantation, Bulford, and Canadian troops occupying sites on the central Plain from mid-October 1914. But new sites were needed and very quickly were chosen south of Swindon and then in the Wylye Valley, where thousands of recruits found themselves sleeping under canvas, wearing their civilian clothes and with no change of garments. At some camps not even tents were immediately available and men had to sleep under hedges.

At first, washing was often done in the nearest river, though several towns made facilities available: from December 1914 to February 1918 the White Hart Hotel in Salisbury provided 13,432 baths for soldiers at 4d a time, which included soap, towel and attendance. (A British private soldier's pay was 1s a day.) A basic menu might consist of boiled beef, potatoes boiled with the earth still on and stewed tea; cooks were usually selected from those men thought to have the least military potential, with little regard paid to their culinary talents. But standard varied from camp to camp; one Wiltshire recruit featured in the newspapers was happy with hot fish or bacon for breakfast, hot meat, beans and potatoes for dinner, and bread, butter and jam or paste for tea – plus 7s a week 'pocket money'. He also received a straw palliasse (mattress) and thick blanket.

After the building of three camps started in the Fovant area, A. G. Street from his farm nearby at Wilton noted the arrival of 'cement, bricks, timber, window-panes, corrugated iron, drainpipes, electric-light plants, surveyors, navvies, bricklayers, carpenters, beer, bad language, and most other ingredients which went to the building of a camp'. Most huts were erected on brick piers, with wooden frameworks and corrugated iron roofs, the planked walls lined with asbestos and heavy paper. There was a door at each end, with six windows down either side. Each hut had a stove and two oil lamps. It was this arrangement that featured in most Wiltshire camps, though huts at Chiseldon had walls of sheet asbestos; they were easy to erect and durable, but damp and cold to live in.

The development of the pre-war camping-grounds of Durrington, Hamilton, Lark Hill and Rollestone required 3,500 construction workers, chiefly carpenters. But by mid-November work was behind schedule because of poor weather and a lack of building workers aggravated by unrestricted army recruitment. (Of 920,000 building tradesmen in Britain in July 1914, 178,000 enlisted in the first year of war.) There was also a shortage of materials, such as seasoned wood and galvanized sheets. The resulting use of unskilled labour and unseasoned wood led to poor quality work.

When war broke out, a carpenter's pay was 7½d an hour; by December this had risen to 10½d. A labourer's peacetime pay of 4d or 5d an hour had increased to 6½d. With Sunday work, a carpenter was receiving £3 a week, a labourer 35s, plus free accommodation and bedding. Recruitment posters were offering single men starting pay in the Army of 7s a week, with 12s 6d separation allowance for married men with no children. When the 7th Wiltshire Regiment arrived at Sutton Veny in the spring of 1915, there was

Building the Wartime Camps

Construction workers at Codford Camp in January 1915.

much discussion and discontent after its members discovered that the civilians building the huts were receiving such high pay; the soldiers thought it grossly unfair that 'shirkers' should be so much better treated than men who had enlisted voluntarily.

Some of the 15,000 or so workers employed by Sir John Jackson Ltd in the Warminster and Lark Hill areas went on strike in February 1915, wanting full pay when unable to work during wet weather, instead of the quarter-day's pay they received if they stood by until breakfast and were still unable to work. Sir John visited the camps and told his men to 'take it or leave it'. Some chose to 'leave it' and quit the district and were replaced; others returned to work.

The choices of some sites, notably those close to the River Wylye and so prone to flooding and mud, proved unfortunate. And, in the rush to provide shelter for the men before winter set in, the authorities decided that only after the huts had been erected would roads be laid. When the weather broke in late October, the result was thick, glutinous mud; between mid-October 1914 and mid-February 1915 it rained on eighty-nine out of 123 days. Mud, certainly in winter time, was to be a feature of most local camps throughout the war. In 1916 a member of the Post Office Rifles recalled in the unit's magazine, the *Castironical*, the winter that had just passed at Fovant:

> We encamped in the inevitable Army mud. The huts were comfortable, but we were subjected to all the usual torture of rain, snow, hail, sleet and frost which go to constitute a typical English winter. We had, however, one consolation. Bad as our conditions were, France must surely be worse.

Building the Wartime Camps

Having spent two days in Wiltshire, a special correspondent for *The Times* commented in the issue of 22 September 1915:

> When the construction of a certain camp, which lies over a hill some distance from the railway was begun, the Government bought up for the purposes of transport a large number of farm wagons. The price paid for the wagons could hardly be described as a competitive one, but they had not been long in use when it was found that the steepness of the road to the camp made horse transport undesirable. Traction engines were then hired for the work at a price 40 per cent higher than would be asked in normal times. These engines drawing heavy loads quickly ruined the road and repeated repairs have cost the local authorities nearly £40,000. Now, when the camp is practically completed, work has been started on a light railway which will connect it with the main line. People are asking why the railway could not have been built at the outset.

The description suggests Fovant Camp, where the military railway opened on 15 October 1915. The correspondent was also scathing about the attitude of the workers. 'Nowhere in the country could the virtues of economy and national organization be more urgently required than in this district,' he complained, continuing:

> if one-half the stories I have heard of idleness and a determination to make an easy and well-paid job last as long as possible are true, then the labour engaged has remained unsatisfactory, and because of its inefficiency has been woefully expensive to the country.
>
> Anger in the villages is all the keener because the worst slackness has been observed among unmarried men of military age. There were recently at work in one locality 3,000 men of whom it was estimated that 1,300 might have enlisted. Among a considerable section of them laziness seems to have been developed into an art, and any display of energy by a newcomer to the gangs has been resented . . .
>
> A farmer upon whose land many of the huts have been erected told me that the men engaged on the job were lazy in the extreme. It was quite a common practice among them to turn up for work an hour and more after the whistle had gone. 'I could not tell you,' he added, 'how many times I have seen them asleep under the hedges when they were supposed to be working. Owing to the scarcity of labour, too, they don't care very much for the foreman. Half-a-dozen men were discovered taking a nap one day, and when they were hauled from the hedge-bottom they said they were tired and had done enough. The matter was overlooked.'
>
> Another farmer told me of two carpenters from Scotland who were heard to say, 'England's a grand place. They give you all day to do an hour's work here,

and you get a handful of money for doing it.' A labourer remarked that the war was the best thing that ever happened and the Kaiser was the best friend they had. A farm hand disagreed with him and knocked him down ...

A number of unmarried youths hanging around an uncompleted hut were asked why they did not enlist. Their reply was, 'Let the married men go and fight. They've got something to fight for and we haven't, and we're not going to chuck up good pay and an easy job.'

A few days later, 'Rusticus', writing to *The Times* from 'Edge of Salisbury Plain', added that men ready to do their ordinary amount of work had been told (by their less industrious colleagues) that if they meant to carry on like that they had better go home. One boy, earning 38s 2d a week, bragged that he didn't do 1s worth of work a day. Another, aged 16, was being paid 48s a week.

The House of Commons was told how an officer had asked a ganger at Warminster how many men he had: 'fifty-five' was the reply, but only thirty-one could be found. Two men appeared in court in April 1915 accused of defrauding the Government by claiming wages for absent or non-existent workers at Codford. They were acquitted, though not before the prosecution had cited a series of stories of fiddling the books.

Some civilian workers put 'OHMS' (On His Majesty's Service) on their vehicles, but the 'War Department' wording and the soldier-driver suggests that this is an Army lorry. The seated man holds a board with 'FARGO 13-10-14' on it, indicating that the photograph was taken at the camping-site of that name near Stonehenge.

Stung by such criticism and to emphasize their contribution to the war effort, construction workers put OHMS ('On His Majesty's Service') on their cars and lorries (which some seemed to think absolved them from registering their vehicles and from observing the speed limit) and on boards they displayed when posing for photographs.

Despite everything, by mid-1915 huts were ready for 850,000 men throughout the country. Most of Wiltshire's new Army camps had been completed, though the introduction of gas warfare led to the establishment of an experimental station at Porton in 1916 and several airfields were yet to be built.

Nevertheless, some soldiers were still exposed to the rigours of camping under canvas because of under-capacity in camps, or because they were based some way away from them. In November 1916 the 2/1st City of London Royal Field Artillery (RFA) found itself in bell tents on the bleakest part of the Plain ½ miles south of Imber. In *Officer and Temporary Gentleman*, the celebrated author Dennis Wheatley recalled:

> The days that followed were sheer purgatory. The rain increased until it was continuous. Day after day it poured in torrents while the men, protected only by their mackintosh ground-sheets, laced around their necks, toiled at digging gun pits. Our chaps had never before been called on to live in the open, so our cooks had no experience of using camp ovens. Petrol and sugar was flung on the fires in vain, time after time wind-driven rain put them out. During lulls they managed to boil up kettles for tea but nearly all our food had to be eaten cold. On that desolate plain there was not a house or barn in which anyone could shelter even for a short while. Drenched to the skin, their boots sodden, unwashed and utterly miserable, the men crouched shivering in their bivvies …
>
> The men began to cough; so did I. Scores of them developed bronchitis, then pneumonia and had to be sent to hospital. The infantry suffered even worse than we did, for two of their men actually died out there on the Plain and their bodies were taken away in small arms wagons. Out of a total force of 498 troops and 16 officers, 170 men had to be evacuated because they were coughing their lungs out.

Criticism of the Camp Constructors

By August 1916 £24.5 million had been spent on hutting for troops, hospital patients and horses in the United Kingdom. That same month the Public Accounts Committee reported critically on the building of the camps, particularly on one company, which it did not name. However, its comments pointed to the company owned by Sir John Jackson, who had offered to build three camps outside Wiltshire for no profit with, it was alleged, the aim of securing other well-paid work. The company had gone on to win contracts for the massive Lark Hill and Codford hutments, for which it received a 5 per cent commission on the actual cost.* The War Office had tried 'cost plus percentage' in 1902–3 and had found it too expensive, but had not put the wartime work out to tender, due to the need to build a great many camps as quickly as possible.

Sir John – the MP for Devonport – was sufficiently offended by the Public Accounts Committee's remarks to ask for a judicial inquiry. A Royal Commission was convened and sat in January 1917. Lawyers for Sir John said he had written to Lord Kitchener on 7 August 1914 offering to build three camps without profit and these were duly erected. For the Wiltshire camps, the firm had charged cost plus 5 per cent, with 1½ per cent being added for establishment charges. Directors' services had been given free. Sir John claimed that normally the added-on profit would have been 10 per cent; in normal times less was never accepted and except for war work the company would ask for 15 per cent. Eventually his company had spent £3.75 million on building various camps, carrying out work that at one time had been estimated at £1.5 million.

The Commission's report in April 1917 exonerated Sir John's company from all imputations, including that of bringing about a state of affairs in which it could, and did, extort exorbitant terms, but added that a state of things had arisen, which enabled him practically to dictate his own terms. The amount he was entitled to under the agreement for the Salisbury Plain camps was greatly excessive and the agreement was unreasonable. The Commission felt that the wages paid were fair, being less than the London rates but more than those hitherto paid locally. Some lazy and incompetent men had been hired, but this the Committee attributed to the need for the huts to be erected as quickly as possible. There was no evidence of excessive prices being charged for materials bought through Sir John's company; indeed they were lower than those paid direct by the War Office.

Sir John's immediate reaction to the report was to write to the War Office reducing his added-on profit claim to 4 per cent, surrendering £30,140. He had made this offer the previous July, but had withdrawn it when the Public Accounts Committee had made its criticism and until the Royal Commission had announced its findings.

---

*The word 'hutment' refers to a group of huts, perhaps a complete camp. It can refer, for example, to the entire Lark Hill complex or just one of the thirty-four groups at Lark Hill itself, each able to accommodate a battalion.

Building the Wartime Camps

Pilfering and Waste

Pilfering at the camps was a common problem, with building materials, blankets, clothing, boots and even mules being popular items, taken either for personal use or for passing on to traders. In 1918 so many Australian boots and leggings were being illegally sold in the Salisbury area that they were almost fashionable. The Ratfyn railway sidings on the line from Amesbury to Lark Hill were a popular spot for thieving, presumably because they were isolated. In mid-1915, after two or three months of losses, Sergeant-Major Ernest Sayer of the Army Service Corps kept watch and apprehended nine labourers helping themselves to stores. At Draycot Camp (soon to be renamed after the nearby village of Chisledon), it was alleged there was 'enough [waste food] to encourage every cottager in half a dozen villages around to start a family of fowls at his door with the certainty of profit in their produce'. It was also claimed that local women got their men to carry bags of soldiers' clothes that they had washed back to the camp because, full of pilfered goods, the bags would have been too heavy for them on the return journey.

The Army's apparent waste of food was a sensitive issue for much of the war. In late 1914 a soldier on active service at home was meant to receive a daily ration of: 20oz of fresh meat or 1lb of bully beef; 20oz of bread or 1lb either of biscuits or flour; 2oz of bacon; 1oz of cheese; 2oz of peas, beans or dried potato; plus tea, jam, sugar, salt, mustard and pepper. A 4d allowance was also made to each soldier to allow him or his catering officer to augment the official menu. In the first weeks of war such rations were seldom available to recruits, but afterwards there often appears to have been food to spare until shortages forced more economical policies.

In February 1915 Sergeant Roberts of Chisledon Camp, giving evidence at the trial of Henry Gardiner for stealing two loaves of bread, admitted that a considerable surplus was thrown away and that 'tons of stuff' were taken from the camp. In June 'at least a score' of loaves of bread were seen floating down the River Nadder at Harnham, close to Salisbury, and were presumed to have come from Army camps upstream, probably those in the Fovant–Hurdcott area. The matter was particularly sensitive, as the price of bread had recently increased to 8½d for a 4lb loaf. On 3 July, a special correspondent from *The Times* reported thus on the incident:

> not many miles up the river lies a military camp which at one time was notorious for its wastefulness. Bread, I was told today, lay about the camp and was trodden underfoot. The farmers are given wagon loads of loaves that had become unfit for human consumption, but is good enough for pigs . . . the camp in question has now a comparatively good reputation . . . It is quite plain to the

men that their ration is a generous one. So it is, and there are a great number of men for whom it is too much.

The reporter claimed that the soldiers left a great deal of meat and vegetables on their plates and that the daily bread ration could be reduced by 4oz to 1lb. Food was provided each day for the camp's entire strength, even if men were absent. It was served up in dixies, and a case was cited of one of these being allocated to seventeen men, thirteen of whom were on leave.

The War Office admitted to a certain amount of waste (though claiming this had been much reduced in recent months), and the worst case had proved to be that of forty loaves, 'not a big percentage' in a battalion of 1,000 men. With reference to the loaves seen in the Nadder, the authorities suggested that troops leaving the district had been given so many delicacies by their relatives that they had jettisoned their Army issue of bread. Some waste was caused 'to a considerable extent by the fastidious tastes of a large proportion of men drawn from a class much above the status of the ordinary Regular soldier'. Some units served all the daily meat ration at dinner, more than many men could eat at one session. Another explanation for the same incident was that a unit leaving the area at daybreak had indented for a full day's bread ration; observing a standing order that storerooms be left clean and empty, its quartermaster, rather than return the bread, had ordered it thrown into the river.

In mid-1915 the War Office announced that the main components of the daily ration per man were now: 1lb of fresh meat; 1lb of bread; 2oz of bacon; and 2oz of sugar. Jam and cheese were omitted, but the cash allowance to be spent on other food to provide variety was now 5½d. But menus varied from camp to camp. When Jacko Thompson of the British 45th Training Reserve Battalion arrived at Sutton Veny in October 1917, his tea consisted of 'a couple of slabs of beef (?) 5 rounds of bread (stale) and a chunk of margarine Also a basin [underlined three times] (+ milk & sugar – which was VILE)'. When he moved to Perham Down Camp his breakfast was of kippers, bread and dripping, with tea to drink. Dinner consisted of steak pie and potatoes and tea was bread and jam, again with tea. Tripe and onions were often on the menu.

Daniel Combes, a Dinton member of the Wiltshire War Agricultural Committee, complained in February 1917 of a local camp 'where pig-feeding on the refuse was quite a big thing'. Censorship prevented the naming of the camp, but it was almost certainly Fovant, 2 miles from Dinton, where in autumn 1915 the 15th West Yorkshire Regiment had its own pig farm in agricultural buildings (as no doubt did other units). Three-quarter loaves of bread were often thrown away; some people were buying it by the cart for pigs and there always seemed plenty of it if they went to the right sergeant, it was alleged. There were also half-legs of mutton and half-sirloins of beef. This was at a time of increasing food shortages and pleas for voluntary rationing; posters would be issued in April exhorting 'Eat Less Bread and Victory is Secure'. A major-general reassured the Committee that there was a monitoring system in place and suggested that the waste referred to by the Committee might have been caused by a newly arrived unit from overseas, which had yet to assimilate instructions.

Despite increasing food shortages, some very special efforts were made to provide first-rate Christmas lunches. In 1917 New Zealanders at Number 7 Camp, Sling, had a choice of halibut, beef, mutton, pork, turkey or goose, with five sweets to follow. An Australian soldier, Ivor Williams of the 21st Battalion, described his Christmas Day at Fovant (or possibly Hurdcott) in 1917:

> We did not get out of bed till 11am. After dressing we went down to the mess for dinner. The tables were gloriously decorated and layed [*sic*]. Cost about 10 pounds to do it and only seat 40 persons. We sat down at 1.30pm. After a nice address by the commandant and waded through an eight course dinner, after which we toasted all to be toasted.
>
> Finally rising from the table at 4.30pm, some going eh? This is not all. At 6pm we started off again and had a three course tea. The manager of the Vaudeville Theatre [at Hurdcott Camp] (he was invited to dinner) reserved the first three rows in his Theatre ... The whole was lovely, so in return we invited the whole Company of Artists to our mess for supper and evening. Supper was served at 9.30pm during which time we had songs and dances, after each course (6 in all). Finally at 1am we all retired after having a glorious day.

Later in the war, great efforts were made to practise 'good housekeeping'. An article in the *Sydney Morning Herald* of 16 January 1918 told of an Australian catering officer's initiative:

> This autumn, when motoring through Wiltshire, he espied a number of orchardists sitting alongside their properties smoking, whose fruit remained unpicked. 'Why don't you pick your plums and apples' he asked. 'There be no sugar,' said the farmers. 'Nobody wants the fruit.' He immediately replied, 'I want it. I'll buy your crops and any others you can put me on.' At 1½d a pound he bought up hundreds of tons of fruit, gladdening the hearts of the Wiltshire farmers. By means of an ample sugar supply he converted the fruit into jam. Hundreds of tons he made, and on every pound he tinned he saved the Australian Imperial Force a little over 5d.

Another Australian, Donald Clarkson, noted that the estates of the Earl of Pembroke around Fovant Camp were teeming with game, as were the rivers with fish, but soldiers were not allowed to catch anything except rabbits (and

Building the Wartime Camps

The butcher's shop at the Australian General Hospital at Sutton Veny looks well stocked and the attendants appear bored, but it is 1919 and both camp and hospital are winding down. Perhaps the men are dreaming of going home.

then only if the soldiers did not damage the undergrowth that acted as coverts for game birds). At least they could pick blackberries and on route marches the soldiers would have a feed of them at every stop.

Units were expected to grow produce in their camps, reporting on their successes (and the quantities of kitchen waste put to good use) in their war diaries. That of the Australian 8th Training Battalion at Hurdcott claimed that:

> good results from our Gardening efforts are assured. Every available Plot between Huts and every spare portion of Parade grounds have been cultivated and sown with Potatoes, Lettuce, Radish, Parsnips, Carrots, Beans and Pump-

kin. For a total expenditure of £65 in seed, fertilizers and Gardening Tools, a conservative estimate gives the value of Vegetables which will be available for the Troops at £350.

The 8th won a silver cup valued at £5 5s for the best display of garden products and for gaining the most points. It even included a photograph of the very impressive display in its war diary.

4
Supporting the Mother Country

The First Canadian Contingent

On the outbreak of war, Canada had only 3,000 Regular soldiers, but by 8 September 35,000 volunteers, many with previous military experience, had arrived for basic training at a hastily built camp at Valcartier, near Quebec. Four weeks later, some 30,617 men, comprising the First Canadian Contingent and attached units, were on their way to England and were joined at sea by 537 men from Newfoundland, then a dominion in its own right. Most ships arrived from 14 October onwards at Plymouth (with one or two others disembarking at Southampton and Avonmouth), from where ninety-two special trains carried the men and equipment to Lavington, Amesbury and Patney & Chirton stations, where they marched to camping-grounds established before the war at West Down North and South, Pond Farm and Bustard.

All the sites consisted only of tents and cooking shelters. A representative of the Canadian Government had approved them while England was enjoying an Indian summer, but had not realized that under the springy turf the hard bed of chalk inhibited drainage. The soldiers soon discovered how inhospitable the Plain could be when they were inspected in pouring rain by Field Marshal Roberts on 24 October. (All most of them saw of the eighty-two-year-old soldier was his car driving past.)

The Canadians faced a thirteen-week programme that should have included an hour's physical training each day, fifteen hours' drill a week and twenty-seven hours' rifle instruction spread over the first three weeks. Night work and entrenching was planned to take place weekly, with progressively longer route marches three times a week. But it was to prove impossible to adhere to this schedule.

The Canadian's British commanding officer, Lieutenant-General Edwin Alderson, set up his headquarters at the Bustard Inn. At first, the troops' canteens were 'dry' (that is, did not sell alcohol), but after thirsty soldiers caused problems in local villages the Canadian Government approved the selling of beer and spirits in camp. Drunkenness caused some soldiers to be returned home, though *The Times* assured its readers that these were mainly men born in the British Isles who had been in Canada for only a few years.

On 2 November, in four hours of pouring rain, the troops held a dress rehearsal for an inspection by George V and Queen Mary two days later. The 4th also dawned wet, but the sun did break through as the King took the royal salute at 2pm, seven hours after the troops had mustered. Newspaper reports were detailed and fervent. One in the *Salisbury Journal* of 7 November described the visit thus:

> Their Majesties' decision to take the first opportunity of personally welcoming the Canadian troops to the Mother Country and thanking them for their noble response to the call to arms at the present crisis in the history of the Empire evoked the enthusiasm of every officer and man in the Canadian contingent.

The royal party arrived by train at Amesbury and went by car to Bustard Camp and on to West Down North and Pond Farm camps, where inspections were held:

> Scenes of great enthusiasm prevailed when their Majesties drove away in their motor car, a picturesque sight being provided by the [Canadian] Highlanders raising their Glengarries [brimless hats] on their bayonets and cheering lustily. The farewell scene was of a very impressive character. For a distance of two miles from Bustard Camp, the Canadians were drawn up on either side of the road and, as the King's motor car left for Amesbury, the hearty cheers which were raised echoed across The Plain and gave ample proof of the Canadians' appreciation of the Royal Visit paid to them.

The King praised the prompt reply of the Canadians to the Empire's call, which he described as of inestimable value both to the fighting strength of the Army and as evidence of the Empire's solidarity. He added: 'I am glad to hear of the earnest and serious spirit which pervades all ranks, for it is only by careful training and leading on the part of the officers, and by efficiency, strict discipline, and co-operation on the part of all that the demands of modern war can be met'.

The poor weather that marked the visit was typical of the winter of 1914–15, with 24in of rain falling in sixteen weeks. Other problems for the Canadians were a lack of professional soldiers among their officers and poor boots that quickly succumbed to the mud. The best footgear was 'rubber boots to the knees', and traders hawked these around the camps at scandalous prices. Someone thought it a good idea to use a snowplough to clear the roads of mud, but this proved 'senseless and useless'.

Despite the wintry conditions, the 16th (Canadian Scottish) Battalion was anxious to retain its kilts and some of its members even appealed to the Black Watch in Scotland for the correct tartan. (As it happened, the Scottish regiments themselves were short of tartan material, a matter that was raised in the House of Commons in early January.) After much debate, known as the 'Battle of the Kilts', officers voted on 21 December by twenty-one votes to seven for the troops to adopt a plain khaki kilt, though one swore he had lived in a Gordon kilt and he would die in one.

Food for the men was said to be good, though Brussels sprouts were the staple vegetable, not the wisest choice for men living closely together; each man was allowed a day's ration of 1lb of bread, 1lb of fresh vegetables and 1lb of fresh meat or bacon, this last being increased by 50 per cent for parties helping to erect huts. A welcome visitor to Bustard Camp in January was Sir Hiram Maxim, inventor of the machine gun bearing his name, who presented the Canadians with 6½ tons of pork and beans in 10,200 cans prepared for immediate use.

As winter approached, civilian contractors were erecting huts for the Canadians at Lark Hill and Sling. They were aided by bricklayers and carpenters among the Canadians, others of whom acted as labourers – all this at the expense of training. But in any case, training was limited. Lewis Harcourt, Britain's Secretary of State for the Colonies, wrote to Kitchener on 16 December to suggest the Canadians were:

> going back on training from that which they received at Valcartier. There they did regular shooting practice and went out for two or three days' manoeuvring under regular service conditions. On Salisbury Plain, they have none of this and, owing to the condition of the soil, they cannot even learn to trench.

Harcourt suggested that the contingent be broken up and trained with the better portions of Britain's New Army – a proposal that would have been completely unacceptable to the Canadian Government. As well as wondering if Alderson was 'quite a strong enough disciplinarian for the job', he also noted that the Newfoundlanders were jealous of the Canadians and were concerned that they might lose their identity; he urged that they be separated as soon as possible and on 7 December they departed for Scotland.

On 9 November the 4th Canadian Infantry Brigade moved from Pond Farm to the new hutments at Sling. By 15 December, 11,000 Canadians were in huts (many at Lark Hill), with another 9,000 expected to join them that week, but this still left 11,000 under canvas, with concern growing for their health and that of their horses. Soldiers were moved into private houses in nearby villages, though the 1st Brigade had to remain camped all winter at Bustard Camp. Stabling was also found for the troops' horses. At Devizes, 900 men and 750 horses were put up, while others found themselves in villages to the north of the Plain. The Canadian Artillery Brigade was billeted at Dauntsey Agricultural School, West Lavington, on 1 January, with the approval of the governors. But the Board of Education, which had an agreement with the

There were severe floods in many parts of Wiltshire in early January 1915, causing much disruption to military activities. These men are on their way from Amesbury to Lark Hill.

War Office about the military use of school buildings, objected and the brigade had to move out on the 8th, much to the indignation of Dauntsey's governors.

Apart from coughs and colds, the Canadians suffered remarkably few serious illnesses when in tents (even if these were of poor-quality canvas and at first without floorboards). Perversely, matters deteriorated after the move into the huts, which were on brick stilts, with the wind blowing up through the floorboards. Of fifty-eight Canadian graves in Wiltshire dating from October 1914 to April 1915, three of the occupants were killed in accidents, six died before 1 December, and forty-seven (many of whom would have moved into huts) between then and 28 February. Causes of death were not always specified, though pneumonia and meningitis were often cited. In early 1915 there was a meningitis epidemic among the Canadians, thought to have originated at Valcartier and to have been exacerbated by overcrowding on board ship and in tents, with twenty-eight cases proving fatal. Of 4,000 Canadians admitted to hospital in a fourteen-week period, 1,249 were suffering from venereal diseases.

Despite their tribulations, the Canadians may have found consolation in the fervour with which the British people, including many in Wiltshire, greeted them – certainly they had the advantage of being the first overseas troops on the scene, at a time when enthusiasm for the war was at its highest. There was detailed and eulogistic newspaper coverage of their activities and sterling qualities, and many patriotic postcards were published. Drawing its information from *The Times*, one postcard caption enthused: 'Nothing like the Canadian Contingent has landed in England since the time of William the Conqueror ... The force has its own engineers, signallers, transport corps, ammunition parks and field hospitals, and there are 34 chaplains and 105 nursing sisters.'

Another postcard, featuring the Canadian camp at Pond Farm, was liberally decorated with flags and included the sentiments: 'The wish of all at Market Lavington [the nearest village] is that God will guard them, and give them victory, and that all will return safe home to Canada' and 'With the Maple Leaf around the Union Jack are we downhearted? No!!!'

Some local residents came to have a touch of self-interest about wishing the Canadians a safe return home, finding some of them uncivilized. Even officers behaved badly in public, twenty-two being among the hundred Canadians arrested for drunkenness one night in Salisbury. This posed a problem to the authorities, who were used to dealing with errancy by the rank and file but not by their superiors. And local girls were not immune to the charms of the Canadians, with liaisons inevitably occurring, especially with many young local males having joined the British Army, There were a few weddings, though in one case, involving a Canadian billeted at Lydeway, the woman turned out to be already married.

Later arrivals from Canada complained of having to live down the reputation of the First Contingent. But it really was news to local people when a report in the *Winnipeg Telegram* on the 'Battle of Amesbury' found its way to

The First Canadian Contingent assembles for a review by Sir William Pitcairn Campbell, General Officer Commanding Southern Command, on 27 November 1914.

Supporting the Mother Country

Wiltshire. It was alleged that thirty Canadians had 'trimmed' 120 British Territorials in a fight that had caused townsfolk to lock themselves in their houses and the 'Mayor of Amesbury' (a dignitary unknown to the townsfolk) to appeal to the military authorities in Salisbury for help.

On 4 February 1915 the King again inspected the Canadians, shortly before they left for France. By now a military railway had been laid from the Amesbury–Bulford line, through Lark Hill to Rolleston Camp, and the royal train steamed along to a temporary station 'close to Bustard Inn', where 25,000 troops were drawn up. As was the custom on such occasions, a royal message was relayed to the troops:

> Officers, non-commissioned officers and men: at the beginning of November I had the pleasure of welcoming to the Mother Country this fine contingent from the Dominion of Canada, and now, after three months' training, I bid you God's speed on your way to assist my army in the field.
>
> I am well aware of the discomforts that you have experienced from the inclement weather and abnormal rain, and I admire the cheerful spirit displayed by all ranks in facing and overcoming all difficulties. From all I have heard, and from what I have been able to see at today's inspection and march past, I am satisfied that you have made good use of your time spent on Salisbury Plain. By your willing and prompt rally to our common flag, you have already earned the gratitude of the Motherland.
>
> By your deeds and achievements on the field of battle, I am confident that you will emulate the example of your fellow countrymen in the South African war, and thus help to secure the triumph of our arms. I shall follow with pride and interest all your movements and I pray that God may bless you and watch over you.

The Canadians started to move to France on 7 February. Left behind in Britain, in the care of the Zoological Society, were four – some say five – bears, the mascots of various units. One became the inspiration for Winnie-the-Pooh, the subject of much-loved stories written for children by A. A. Milne. Shortly after the outbreak of the war, Lieutenant Harry Colebourn was travelling by train to Valcartier to join the First Canadian Contingent. He changed trains at White River Bend, where he saw a trapper with a female bear cub whose mother had been shot. Colebourn bought the cub for $20, naming her Winnie, after Winnipeg, his home town. On Salisbury Plain she became the soldiers' unofficial mascot and slept in Colebourn's tent. On his way to France on 9 December 1914, he left Winnie at London Zoo (home of the Zoological Society), where she became a firm favourite with visitors and keepers, being very tame and well behaved. After the war, Colebourn went to reclaim Winnie, but decided to leave her there after seeing how well loved she was. In 1924 A. A. Milne visited the zoo with his son, Christopher Robin,

Supporting the Mother Country

The original Winnie-the-Pooh, seen here with Harry Colebourn, the officer who adopted her in Canada and brought her to Salisbury Plain.

who became enchanted with Winnie. Two years later, Milne published the first Winnie-the-Pooh book.

Given the poor weather, lack of huts and limited training, it is remarkable how during its time on the Plain the First Canadian Contingent evolved from a hastily assembled force of 30,000 officers and men, some of whose members had little or no military experience, into the First Canadian Division of 18,000. (Various units were hived off and some officers and men were transferred to British units.) Nearly all of their equipment had to be replaced and augmented from British supplies. Hundreds of wagons were discarded, and even more useless were 25,000 entrenching tools that were later sold for scrap. As late as 23 January, General Alderson was noting such inadequacies as: no travelling field-kitchens; only two machine guns to a regiment; infantry lacking any telephones; and only forty-eight out of 600 replacement bicycles to hand, the Canadian-pattern machines having been condemned.

Rough, but not ready, the original Contingent may have been, but the First Division was soon to distinguish itself in action, at the Second Battle of Ypres, where its fortitude repulsed German attacks and saved the day. During its time in the Front Line from 15 April to 3 May, its killed, wounded and missing numbered 208 officers and 5,828 other ranks. Three Victoria Crosses were won.

Australian Soldiers

When an Australian officer visited Salisbury Plain in October 1914 he found nothing yet had been done to build huts for a convoy of troops on its way from Australia and New Zealand. An Army doctor reported that to 'house Australian troops in tents in mid-winter on this windswept area after [a] long voyage in troopships passing through tropics and sub-tropics would be criminal'. (In August 1917 the Australian commandant at Perham Down Camp was to make the point that Wiltshire winters endangered the health of his men.) Representations to Lord Kitchener resulted in the troops heading for England being diverted to Egypt and then to Gallipoli, where their massive casualties suggest they would have been better off on Salisbury Plain.

From June 1916 men of the Australian Imperial Force were based at most of the Plain's camps, where they trained to provide reinforcements for divisions at the Front; many had already seen active service and were recuperating from wounds and sickness. Initially, quarters were allocated for 43,000 Australians, though this number was to grow and they would be the predominant nationality on Salisbury Plain for the rest of the war. The Australian Imperial Force (AIF) headquarters in the United Kingdom were at Bhurtpore Barracks at Tidworth.

On 27 September 1916, George V reviewed Australian and New Zealand troops at Bulford Field, one of several royal inspections of ANZAC troops in Wiltshire. The Australian Major-General John Monash remarked that 'apart from being the biggest and most splendid, it was much the most successful review I have ever been present at'. Afterwards Monash delivered the royal party back to its train one minute before it was due to leave. George V thought it 'splendid timing' and stated that Monash's was 'a very fine division. I don't know that I've seen a finer one'.

Some people in the Warminster area viewed the Australians' arrival with trepidation, and there was talk that the town should be put out of bounds to them. One lady warned that 'unless they get what they want they break everything and everyone that come in their way'. An Australian wrote to the *Warminster Journal* complaining of comments such as 'they are rotters' and 'not fit for decent people to mix with'. He asked for the Australians to be judged by their actions and this most people seemed willing to do, for on arrival they were given a civic welcome. At Wilton a particular devotee of the Australians was Miss Uphill, whose refreshment room for them in South Street became known as 'Miss Uphill's Dugout'. She wrote to the *Brisbane Daily Mail* 'to say how much the Australians enjoyed the beauty of the local countryside and to remark on how they fussed over children and were courteous to women'.

Early in the Australians' stay in the Warminster area there was a minor fuss because some of them were visiting several restaurants for a succession of free teas and there was drunkenness and other minor offences, but the town survived regular weekend incursions by hundreds of Australians – not that they had a particularly good reputation for military discipline. Being away from camp without permission was a common offence: in June 1917 there were 2,800 other ranks at Perham Down, but that month 427 men had been charged with 'absence without leave' (AWL), with forty more accused of 'breaking camp'. All told, the Australian Provost Corps based at Tidworth dealt with 25,890 AWL cases throughout the country in 1917 and 1918. Also frequently noted were cases of 'failure to salute', even officers being among the offenders. In November 1917 the Australian commandant at Sutton Veny conceded that:

> the worst forms of possible crimes have been committed by members of the A.I.F. . . . amongst the troops in this district. There is no doubt that a gang of blackguards is at work among the various Depots . . . such things as sandbagging [mugging] and robbery amongst the soldiers have been rampant, but also stealing, robbery and begging among the civil population . . . The perpetrators of these outrages are probably men who in civil life are gaolbirds . . . who in many cases have spent the greater proportion of their soldiering days in detention, or absent without leave or in venereal Hospitals.

The Reverend George Atwood of Bishopstrow, just a mile from Sutton Veny, made frequent complaints about Australians, prompting the provosts' commanding officer to note that 'he always puts the worst construction on any rumour that he may chance to hear to their detriment'.

There was trafficking in Army equipment and local magistrates' courts dealt with many cases of civilians being in possession of items of Australian kit, the usual fine being a pound or two. In January 1918 Mrs Maguire of 50 Pound Street, Warminster – the wife of an Australian soldier – was found to have forty-seven items in her house. Sometimes soldiers were arrested with bags of clothing intended for sale elsewhere in the country and stolen goods also found their way to local shops, which from time to time were raided by the Australian and British authorities. Many farmers reported minor thefts of animals, birds and tools, but often the provosts were unable to find evidence that the crimes had been committed by Australians. Cases of trespass were also common, Charles Penruddocke of Compton Estate west of Salisbury being a frequent complainant. But when the provosts' commanding officer visited the Sutton Veny area in October 1918, he reported that the civil police, farmers and gentlemen to whom he had spoken were decidedly 'in favour of the conduct of Australian troops'.

It is said that in pubs in the Warminster area Australians used pound notes as spills to light cigarettes, thus demonstrating they were better paid than British troops. One of several serious incidents around the time of the Armistice happened at Bulford; writing home on 12 November 1918 (the day after the Armistice was announced, leading to much celebration), Corporal William Beer (an Englishman in the Royal Army Service Corps) reported:

> The Australians who are in hospital here broke away from their armed guard & invaded the 'Wet' canteen where all the beer is sold. These men are not allowed to drink beer while in hospital with the result that ... they simply soaked themselves in it & then the quarrels began.
>
> An armed guard of about 6 men were brought on the scene but the Australians overpowered & disarmed them, & then smashed everything within reach, every window in the canteen was broken, they emptied the till & smashed the piano to atoms.

Eventually the rioters were confronted by British troops armed with sticks and iron bars and were gradually brought to order. Beer refrained from telling his family that the hospital treated venereal diseases under a regime that was close to that of a prison's.

A. G. Street probably did not embellish the facts in *The Gentlemen of the Party*, his fictional account of life in the Fovant area, when he wrote of the Australians' behaviour in 1919:

> For discipline and regulations they care not a jot. They fight the military police. Some of the worst characters deserted their regiments and lived rough in the adjacent woods in far worse than Robin Hood fashion. While the main body remaining in camp jeered at their officers, insisted on lifts to Salisbury from every passing car, and made themselves a general nuisance to the world around them, renegades went in for robbery under arms, even murder. For [the] district the period was an awful one, much worse than the war.

There were tales of the men hiding in the woods 'sandbagging' victims, something recalled by Bob Combes of East Farm, Fovant, though claims noted by

Donald Clarkson that two soldiers from Hurdcott had been killed are unlikely and no such reports have been found in the war diary of the Australian Provost Corps.

It is difficult to assess quite how much of a problem the Australians proved to be and whether they were worse than other nationalities. For almost three years they were the most represented nation on the Plain, which inevitably meant that they committed the most crimes, and in 1919 many spent the last weeks of their stay in Europe in Wiltshire camps becoming more and more frustrated with the delays in getting home. Reports of the Australian Provost Corps are more readily available than any that might relate to British, Canadians and New Zealand troops, and each weekly returns lists a small number of misdemeanours – not just those of soldiers, but also of civilians pilfering, wearing stolen clothing and offering sexual services. The First Canadian Contingent had its own reputation for poor discipline, but it was only in Wiltshire for four months and its camps were somewhat isolated,

Australians at Codford in May 1919 wait for a ship home. They have signed their names on the back of this card and their service records show that one had won the Military Medal for rescuing a wounded comrade under fire and another had spent seventy days in hospital with VD. Many of the men shown here left England on HT (Hired Transport) Main *on 23 July. Another opted to travel home at his own expense, but his address for part of 1919 was c/o an unmarried lady in London, so perhaps he remained in England.*

whereas many of the Australian hutments were very close to villages whose residents must have felt overwhelmed by thousands of men. As the Australian authorities often noted, the vast majority of troops were well behaved and many misdemeanours were minor. The most frightening misbehaviour, occasionally amounting to riots, took place in camp.

After the war ended, Australian troops who had returned from France lingered in many of the Wiltshire camps awaiting repatriation. An early aim was to return home an average of 500 men a day and by mid-March 1919, 4,000 to 5,000 troops were embarking weekly, with the total for that month being 15,842. A shortage of shipping caused delays and led to increasing frustration and unruliness, exacerbated by the continuance of military discipline, which was alleviated only in small part by the setting up of touring cricket teams and vocational training.

From his Tidworth base Major-General Sir Charles Rosenthal, General Officer Commanding Australian depots in the United Kingdom, made many visits to camps to discuss complaints and suspicions about the delays, at one stage having to refute rumours that ships intended for his soldiers had been allocated to other nationalities. Some officers on leave who were telegraphed notifications of berths were found not to be at the addresses they had left at camp, thereby literally missing the boat. In some cases 'apparent glaring neglect of duty has been disclosed', declared the Tidworth headquarters, warning that general courts-martial were being considered.

The Australians left their mark on the hillside at Compton Chamberlayne, where they carved a map of Australia, and at Fovant and Codford, where they cut so-called 'rising suns' – actually trophies of swords and bayonets surmounting a crown. Some men never returned home but are buried in local churchyards, victims of the Spanish influenza that killed millions worldwide in 1918–19.

New Zealand

New Zealand may have sent a smaller contingent to Britain than Australia and Canada, but it was better prepared for war, thanks to 'universal training' (national service), which meant a well-drilled and disciplined force from the start. Further, the Great War had started during the Army camping season in New Zealand, with some 30,000 troops in a state of readiness and such was their eagerness to take part that the 10,000 men comprising the expeditionary force had to be selected by ballot.

Richmond Park in London was suggested as a suitable camping-ground, but New Zealand's senior officer attached to the War Office, Major George Richardson, recommended a locality away from the attractions of the city.

Salisbury Plain was chosen and by 7 October 250 New Zealanders who had enlisted in Britain were sent to Sling under Captain F. H. Lampen, their main task being to build hutments for the main contingent from home. The camp-builders included five All Blacks, an English rugby international and three Rhodes scholars from Oxford – and enough licensed surveyors for an instructor in field sketching almost to find he was the one being instructed! They lived under canvas, though they did have an 'amusement hall' painted in imitation of a Maori dwelling, with a Maori god shown chewing with relish a German eagle. When it was decided that troops from New Zealand and Australian troops were to land in Egypt rather than in England the men left for the same destination on 12 December.

For the next year, most New Zealanders in Britain were convalescents, but as more and more soldiers arrived in Europe it was decided in January 1916 to take over Hornchurch as a general depot for all New Zealand soldiers in Britain. A few months later Hornchurch was made exclusively a convalescent centre to which all those leaving hospital were sent. On 17 July 1916 Codford was selected as a command depot to make fit for active service those who had convalesced. By the end of the year those who had been wounded or were ill were taken to New Zealand General Hospitals at Brockenhurst or Walton-on-Thames, or, if these had insufficient places, to a British military hospital. From there, many were sent to Hornchurch, where their fitness was assessed. Soldiers not likely to be fit within six months were returned to New Zealand; those deemed fit for service or likely to become so within six months were transferred to the command depot at Codford. After fourteen days there, men fit for general service were sent to Sling Camp, next to Bulford Camp. Others were given increasingly rigorous training until passed fit and then sent to Sling, which also trained reinforcements newly arrived from New Zealand before they went on to France. By September 1917 Sling, designed to accommodate 4,000 men, was holding 4,500, and further New Zealand camps were built elsewhere in the country.

At Sling, newcomers from New Zealand were given a short time to recover from the inactivity of their eight-week voyage before starting a ruthless training regime. They had to remove their New Zealand reinforcement badges, which were regarded as the privilege of those who had been on active service. NCOs straight from home lost a stripe and had to pass an end-of-course exam to retain even their reduced rank; it was pointed out to them that there were many privates with more experience serving in France and still awaiting promotion. Likewise, warrant officers were reduced to corporals. (Similar measures applied to Australians.)

At first, training for new arrivals lasted from first light to 5pm, but later this extended to 9.30pm, with night parades and exercises. One feature at least two days a week at New Zealand camps was 'Piccadilly', with all squads

New Zealand troops, including several Maoris, wait for dinner at Sling Camp on 25 June 1918. Four are wearing the New Zealanders' distinctive 'lemon-squeezer' hats.

assembling at 12.25 and, headed by a model platoon of war-hardened soldiers from France, marching past the camp commandant.

When the 28th Reinforcement arrived at Sling on 24 September 1917, it included several conscientious objectors who had been required to accompany their fellow countrymen from home. As part of their objections to wearing uniforms they refused to put their boots on, but were 'persuaded' to walk into camp without them. Archibald Baxter devoted a chapter in his memoirs, *We Will Not Cease*, to his treatment at Sling Camp where, on arrival, he was put in handcuffs 'until he promises to obey'. Thus manacled, he was exercised in front of soldiers to humiliate him and was forcibly dressed in uniform. However, some of his guards were sympathetic, one sergeant even offering to take him down to a girl he knew, but not just for a social call – an offer that was refused. Baxter tells how, when he was sent to France, the guards came up to shake his hand and cheer him on his way. (In France, Baxter's refusal to obey orders earned him repeated sentences of Field Punishment Number 1, which included being tied to a post for several hours a day in all weathers. Eventually he was diagnosed as insane and was sent home in August 1918.)

New Zealanders awaiting demobilization at Sling in 1919 became particularly frustrated with the discipline and training that continued as if the war was still on, and there was 'trouble' in the camp. After their plea for discipline

The caption on this postcard proclaims that the Kiwi above Sling Camp is 420ft high, its bill is 65ft long and the letters are 65ft high.

to be relaxed was rejected, soldiers looted the canteen and officers' mess. To quieten them down, their officers promised there would be no repercussions – then arrested the ringleaders, jailed them and shipped them home.

To keep the others occupied, they were put to work carving the shape of a kiwi on a nearby hillside. Sergeant-Major Percy Blenkarne, a drawing instructor in the New Zealand Army Education Corps, visited the Natural History Museum in London to obtain the bird's dimensions; then, under the supervision of Captain A. M. Clark and other New Zealand engineers, fatigue parties from the Canterbury Battalion were put to work removing 12in of topsoil and replacing it with chalk. Work began in February 1919 and was finished the following month. 'The Kiwi' can still be seen (map reference 199439), though it has lost the 'N Z' initials of the original design.

Other Nationalities

Soldiers from many other nationalities visited Wiltshire to use and inspect its training facilities. Japanese officers were at Sutton Veny in late 1915 to see how British soldiers were trained and South African troops were noted at Chisledon and Perham Down in 1919. Representatives of Allied armies would have attended artillery schools and the specialist chemical and trench training facilities at Porton.

It was the airfields that attracted the most cosmopolitan array of students, continuing a tradition started before the war at the Bristol Flying School at

Lark Hill, where there had been would-be pilots from Australia, China, Italy, New Zealand and other countries. Three officers of the Imperial Japanese Navy were due to visit Netheravon Airfield in September 1913. Two officers of neutral Norway's Flying Corps were at the Central Flying School, Upavon, in 1917, as were Russians, who were also noted at Yatesbury Airfield.

In 1916 Russian artillerymen were at the School of Gunnery at Lark Hill and in April Colonel Novogrebelsky reported on trials they had carried out there using British howitzers' 4.5in ammunition. He noted very weak destructive action of high-explosive shell and unsatisfactory action of high-explosive and time fuses. The Ministry of Munitions indignantly asked 'are we to accept the casual report of a Russian Colonel as a significant condemnation of what has hitherto been regarded as one of the most serviceable and effective weapons supplied to our Armies?'

Few Americans have been noted in Wiltshire during the Great War (their major English camps being east of Winchester in Hampshire, some 24 miles from Salisbury), but several were doctors at British camps, some were members of the First Canadian Contingent and others joined the British Army, including the Royal Flying Corps (RFC). Several hundred members of the Aviation Section, US Signals Corps (which became the US Army Air Service in May 1918) trained at the county's airfields to the extent that they had their own Young Men's Christian Association (YMCA) facilities. Two American lieutenants spent part of 1918 at Porton Experimental Station expanding on research they had done in the States. Personnel of the 154th US Army Aero Squadron were at Lark Hill (presumably in transit) from 9 to 19 March 1918. An American soldier was involved in a disturbance at the George Hotel, Codford, in November 1918; he was probably from his country's Services of Supply Base in the locality. Such a unit would appear to have been relatively small; certainly not large enough to justify the idea of Codford Camp being vacated by its Australian residents, as had been suggested in August 1918. Number 26 (South African) Squadron was formed from members of the South African Aviation Corps at Netheravon on 8 October 1915, moving out on 23 December to Mombasa in East Africa.

5
Rail, Road and Air

Railways

Railways played a significant part in the military history of Wiltshire, for many years being the major way of transporting large bodies of troops. An early example of this role occurred in May 1867, when a Volunteer Review was held near Salisbury of military units from Wiltshire, Hampshire, the Isle of Wight, Dorset, Gloucestershire and Somerset. Several railway companies cooperated in conveying troops and running excursion trains for spectators. When choosing a new manoeuvres area, the War Office must have noted that one advantage of the Plain was that railway lines ran around all its four sides, with a number of stations – albeit some very small – facilitating the offloading of men and supplies.

The 1867 exercise was modest compared with the moving of soldiers to many parts of the country after the manoeuvres of September 1898. Transporting them and 600 tons of military equipment to Wessex was relatively simple, as there had been a steady build-up of troops on the Plain for several weeks and the two opposing 'armies' assembled some way apart, in the Wareham and Salisbury areas. But after the manoeuvres the participants massed near Amesbury to be dispersed over two days, during which 130 trains containing troops and thirty-two bearing hired transport vehicles ran from Ludgershall, Weyhill, Salisbury, Milford Goods Station (Salisbury), Porton, Grateley and Hungerford. Between 7.30pm on 8 September and 9.10pm on the 10th, tiny Porton Station alone handled thirty-five battalions in thirty-one trains, as well as selling 6,000 train tickets to spectators returning from the march past on Boscombe Down. At Ludgershall Station, serving a village of 500 souls, a 1,000yd platform had been installed, being divided into four loading berths. Thirteen railway companies sent carriages there to convey the troops home in thirty-six trains.

In addition to the stations mentioned above, those of Wishford, Lavington and Patney & Chirton saw much military traffic in the pre-war years, with sidings added to cope with the new business. During the summer, details of stations nearest to the various camping-grounds were even given in *The Times* for the convenience of civilian visitors, though no information was given on how to get from them to the camps, often several miles distant. When the newspaper recommended Wishford as the best station for West Down Camp (8 miles away), a subsequent issue cautioned that it was only a 'village siding' (even so, it was used to detrain troops destined for West Down in the 1899 manoeuvres). Soldiers themselves were often faced with long marches to and from their camps, and Perham Down and Windmill Hill were popular with Volunteers and Territorials (whose civilian occupations seldom equipped them for much marching), because they were only a short way from Ludgershall Station.

New lines penetrating the Plain's heartland were required, though the War Office successfully opposed a 1897 proposal by the Great Western Railway to link Salisbury and Pewsey with one running through the Avon Valley, which would have 'destroyed' the value of the River Avon in providing troops with practice in crossing rivers. Several other schemes, including one for a line between Grateley and Lavington, also came to nothing.

The first line to be built for military reasons was the 2.4-mile Tidworth branch line, completed in 1901 and extending from the Midland & South Western Junction Railway station at Ludgershall. The station at Tidworth was a civilian one, though one of its three platforms was reserved for troops. From the station through the barracks ran the Tidworth Military Railway (though there was a tendency to regard the entire track from Ludgershall as this). The line was used to convey materials for building Tidworth Barracks and then to transport regular supplies of food, coal and military stores, as well as troops. Outgoing trains carried manure from the barracks' stables for the Hampshire strawberry fields. So busy became Tidworth Station that its receipts totalled more than those for all other MSWJR stations combined. Nearly all the line between Tidworth and Ludgershall closed in 1963, some years after the track through the barracks had been lifted.

Not strictly a military railway, but owing its brief existence to the building of Tidworth Barracks 7 miles to the south, was the 1.75-mile line opened in 1902 from Grafton to Dodsdown, where a works was established to provide bricks for the barracks. Three transfer sidings were installed at Grafton Station. The line closed in 1910. A mile from Grafton was Wolf Hall Junction, where the MSWJR crossed the GWR; when the War Office sought improvements to this, the two companies shared the £1,000 costs, the new arrangements coming into use on 28 July 1902.

In 1898 the War Office approved the London & South Western Railway's proposal to build the Amesbury & Military Camp Railway running from the Basingstoke–Salisbury line near Newton Tony to Amesbury and on to Shrewton, with the possibility of an extension to Tilshead – which would have been very handy for transporting troops and supplies to the West Down camps. After some groundwork had been completed west of Amesbury, the Army realized that a line past that point would interfere with manoeuvres, so the branch terminated after 4.5 miles at Amesbury, where the station opened in 1902; it included a military platform and four military sidings. The line was extended 1.7 miles to a civilian station at Bulford in 1906 and then 1.1 miles through Bulford Camp, where there was a platform and loop, to a terminus at Sling. Nine trains a day then ran from Amesbury to Bulford. During the Great War a small siding for wagons carrying forage was added, as was one leading to the camp butchery and bakery. The war also saw priority given to military traffic at the expense of civilian passengers and soldiers going on leave. A. G. Richardson had the 'most awful night I ever spent' at Amesbury Station in February 1915, when he waited from 6pm to 1.20am for a train, which finally set off at 4.40, arriving in London at 9am. Significantly in 1918 it was advised that 'to save time' goods for Netheravon Airfield should be sent to Pewsey Station 10 miles away on the relatively uncongested GWR, rather than to Bulford, just 5 miles distant. On the other hand, pre-war Southern Command standing orders counselled soldiers to travel by the LSWR to Amesbury if they wished transport to West Down and Pond Farm camps (as well as those nearer to Amesbury). The orders acknowledged that Lavington on the GWR was closer but warned that transport could not be arranged from there.

The terminal points of the Bulford and Tidworth branches were only 3 miles apart and a 1904 Railway Clearing House atlas shows a projected link between the two. This never materialized, although in May 1907 Richard

A scene of much activity as troops and their equipment arrive at Amesbury station.

Haldane, the Secretary of State for War, admitted in the House of Commons that in bad weather it was preferable to take the 'high road' route of 8 miles between the two barracks rather the 4-mile trackway connecting them; he noted that the journey by train (via Andover) was 26 miles. A rail connection would have been easy to lay over flat ground, but instead the trackway was improved; nevertheless in 1916 Francis Brett Young described his trip as a passenger in an ambulance between Tidworth and Bulford as being 'over the cruellest roads in the kingdom'.

In July 1909 a military platform 650ft long and 18ft wide opened at Patney & Chirton Station, in time for massive troop movements over the weekend of the August Bank Holiday. On Sunday, 1 August, the GWR and LSWR moved 29,000 men, 6,000 horses, ninety field guns, thirty-six machine guns and 130 ammunition wagons in 140 trains, as well as the usual civilian Bank Holiday traffic. Trains left Paddington at forty-minute intervals for Lavington and Patney & Chirton stations, where 100 GWR staff helped with the detraining of 15,000 men, 3,400 horses, forty guns, thirty machine guns and seventy-five carts and wagons of the 2nd London Territorial Division. Trains arrived from 3.30am at forty-minute intervals, with only two being late. At Patney & Chirton the new platform proved invaluable, enabling each train to be allowed thirty-five minutes to unload and clear the station. Lavington Station was suitable only for infantry, having a narrow wooden platform 200ft long and few facilities for unloading carriages and horses.

When war broke out, further sidings were built at many Wiltshire stations to cope with increased military traffic and several new lines were laid to camps, initially so that construction materials could be delivered but subsequently to transport stores and men.

On 2 September 1914 the MSWJR was asked to install a 500yd siding off its line between Swindon and Marlborough, running to the planned site of Chisledon Camp. It was ready for use by the 30th. Then a line was built running from Chisledon Station to a wooden platform near the camp hospital. Later the platform was replaced by a double-sided concrete one that can still be discerned (map reference 190777). The line saw much service in 1919 when the camp was a demobilization centre, but was lifted soon after.

The GWR spent £574 on building sidings at Banwell and Sandford, near Weston-super-Mare, so that stone could be transported from there to camps being built by Sir John Jackson 'at Salisbury' (presumably those at Lark Hill). By 11 December 1914, the railway company was carrying out work required by the War Office at Wylye, Heytesbury, Codford, Warminster and Westbury stations. After the Armistice the GWR listed all the projects, their costs (which had been reimbursed by the Government) and assessments of whether or not the improvements would benefit the company in peacetime. The costs of work at the five stations just mentioned were £1,404, £1,715, £7,812, £2,801 and £9,642 respectively. At Lavington £610 had been spent and the GWR had also carried out work costing £4,916 on sidings at Tidworth, presumably on behalf of the MSWJR whose station it was.

Wylye, Heytesbury, Codford and Warminster stations were very convenient for local camps, but it is a little surprising to read the comments in *The Times* of 29 September 1923 by Lord Long of Wraxall following the death of Sir Henry Sclater. He said that when Sir Henry was Commander-in-Chief Southern Command (from March 1916 to 1919):

> Westbury Station, which is the one I use, was the scene of most of the arrivals and entrainments of troops and recruits for the Salisbury Plain area, and the station-master, a very capable man, told me that previous to Sir Henry's taking over the Command he had found the work almost impossible to perform, so great was the confusion, but . . . 'the new General hadn't arrived a week before everything was altered, and an admirable system devised, for now I can do double the work with more efficiency and with less anxiety and fatigue than was possible before he came'.

It is very difficult to accept this claim of the station's prime importance as there were no Army camps close to it and at least five others would have handled far more equipment and men. But it would have seen a great deal of military rail traffic as a junction, with trains passing through to and from local stations serving camps in the Warminster area. Indeed, the Westbury stationmaster was one of several commended for their efficiency when in June 1916 eighty-eight GWR trains moved the 60th Division from Warminster and Codford.*

At Heytesbury and Codford the GWR's work and costs did not include the building of substantial camp railways, which, as at Lark Hill, was done by private contractors. From Heytesbury Station ran the Sutton Veny Camp Railway. Laid late in 1914, it extended beyond Sutton Veny Camp to Sand Hill Camp. There were 4 miles of track, a quarter of which were sidings.** Ambulance trains ran alongside the reception area of the Sutton Veny Camp hospital so that wounded men could be transferred to wards without delay (though in 1919 Australian patients were taken to Warminster Station by

---

*Westbury seems to have had a relatively quiet war. When Geoffrey Smith and Robert Gilson (close friends of J. R. R. Tolkien, later to be famous for fantasy novels such as *The Lord of the Rings*) visited it in September 1915 they found it to be 'almost without soldiers'; indeed, little of military importance has been noted about it then – though it did boast 'Couzen's Military Stores', and in 1914 300 officers and men of the Army Service Corps were billeted there over Christmas, with the prospect of another 1,000 men and 100 horses being accommodated in the neighbourhood. At the end of the war an RAF stores distribution park was being built near Westbury Station.

**Varying lengths of track (often stated to the exact yard) have been recorded for several camp railways, even within railway company records, and it may be that the discrepancies are due to modifications of the system as the war progressed.

road on their way home). Within the camp complex, stores were distributed over a 2ft-gauge track. There were at least two incidents on the Sutton Veny railway involving runaway trucks. A gateman was killed in late 1915 and early next year soldiers took the brakes off trucks standing at the top of an incline close to Number 6 Camp; two of the trucks crashed through the level crossing on the Sutton Veny main road, coming to a stop at Heytesbury Station. The line was removed in 1923.

Running from Codford Station, a 2.4-mile branch with 1.3 miles of sidings served the camp, opening in 1914 after being constructed in three weeks. It extended past the camp hospital to the outlying hutments on the Chitterne Road, with a spur passing on the north side of St Mary's Church to Camp Number 5. It closed on 1 January 1923, the track being removed by 1925. On 14 March 1915, an agreement was signed by the GWR and Sir John Jackson's construction company for the former to provide workmen's trains between Salisbury and Warminster, one day in each direction, calling at all stations and Upton Lovel Crossing, where a siding had been added in November 1914. (Lovel was then often spelt with one 'l'.) The contractor was to provide metal discs inscribed 'Workmen's train Salisbury Codford to Warminster' and would pay the GWR 'in respect of any works incidental ... to the use of the Crossing for purposes of the said train and the access to and from the same to and from the public roadway', and also the expenses of a uniform, stores and wages 'for a man to supervise the entraining and detraining of workmen to the said level crossing and of a signalman and other workman ... to protect the said crossing'.

The Lark Hill Military Railway was laid in 1914–15 to assist in the construction of hutments and was the most complex of the Wiltshire military lines, with many platforms, loops and sidings and a total of 10.75 miles of track. From the Amesbury–Bulford line it crossed the River Avon by a viaduct, on the western side of which was Ratfyn depot, where there were three reception sidings (actually loops) for wagons. Then it ran through Lark Hill to Rollestone Camp. Very early on, a short-lived spur, the 'Flying Shed Branch', led to hangars until recently used by the Bristol Flying School; camp construction workers slept there and the sheds may have been the delivery point for building materials. For a while Rollestone was the terminus for the line before it was extended past Fargo Hospital to the airfields near Stonehenge and Lake Down. A 750yd tramway linked with the railway west of Lark Hill and served the Hamilton Battery of heavy guns. On 4 February 1915 a special train carried George V, Queen Mary and Lord Kitchener to review the First Canadian Division before it departed for France. Thereafter traffic appears to have been almost entirely stores and material, the line being unsuitable for troop trains, though on Saturdays in the early 1920s a very small train ran from Lark Hill to Amesbury, where its one coach was attached to the Salisbury

Rail, Road and Air

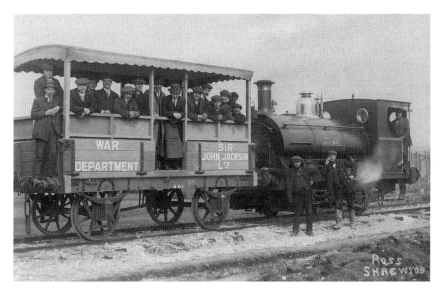

This locomotive and wagon were part of the royal train that conveyed George V and Queen Mary (in their own carriage) on the Lark Hill Military Railway on 4 February 1915. Normally they transported workers and supplies for the camps.

service. Track serving the airfields was lifted in the early 1920s, the remainder of the line closing by 1929.

In 1917 two Canadian companies of engineers laid tramways on the Chapperton Down Artillery Range, one branch leading to Middle Barn on the Tilshead–Chitterne road.

A mile north of the Bustard Inn, a 'tank practice railway' ran close to Shrewton Folly (a copse, not a structure) and into Blackball Firs. Of 2ft 6in gauge, it carried an unmanned 'locomotive' consisting of an engine on a chassis and a trolley carrying a tank-shaped screen at which artillery fired. It is said to have opened in 1916, but tanks were a British 'secret' weapon until September that year and the Germans did not launch their own versions until early 1918. One wonders if the British Army would have been sufficiently ahead of the game to have provided anti-tank training in 1916 and the railway does not feature in a large-scale map of 1924 (nor, indeed, one of 1939). By 1960 it had fallen into disuse.

The Fovant Military Railway opened in October 1915 and ran 3 miles from Dinton Station on the Salisbury–Exeter line to the new camp at Fovant. Unlike other camp railways it was built after the hutments had been mostly completed, when it was discovered that road vehicles could not cope with the local hills; the railway itself had a gradient of 1 in 35. On 28 October of that month the

LSWR approved additional sidings for the War Office at Dinton and referred to its engineering committee a proposal that a junction to the camp railway be 'laid in' with the sidings. In November 1915 the company decided to enlarge the booking office at Dinton, reflecting the increased number of passengers.

During 1919 Fovant was a major demobilization centre handling large numbers of men, so the line saw many troop trains and a regular passenger service using LSWR carriages. A military timetable for June 1919 shows ten trains a day from Fovant to Dinton and nine in the reverse direction. Ten minutes were allowed for the journey from Fovant, with an extra five minutes for the return trip with its adverse gradient. After demobilization had all but ceased in January 1920, the LSWR ran a twice-weekly freight train from April to November of that year, probably to remove surplus materials. The line closed in December 1920, only to be opened again from 1921 to 1924 for the disposal of road stone (presumably rubble from demolished camp buildings), after which the track was lifted.

At Porton Experimental Ground a 2ft-gauge light railway was started in 1917 to link North and South Camps; it was then decided to extend the line to Porton Station on the LSWR, though work was not completed by the Armistice, when 8 miles of track had been laid and there were five locomotives and 150 wagons. The southern part of the line was lifted in the early 1920s, when a tramway was laid east of Winterbourne Gunner across a trench mortar and howitzer range to Thorny Down. Part of the original light railway continued in use until the 1950s.

Boscombe Down Aerodrome was served by a 530yd siding off the Amesbury & Military Branch. It was built in the winter of 1917–18 and was removed after the war.

During the war the LSWR built a siding for the loading of timber at Wilton. For the same reason, a siding and platform were built on the GWR's Swindon–Highworth branch near Stanton Station, where Canadian forestry battalions were cutting trees. Trains of thirty-five or forty wagons would haul the timber away on Sundays. There were extensive sidings off the Highworth branch serving the ammonium nitrate factory at Stratton.

At Ridge Quarry ammunition depot there was a 2ft gauge track within the workings and a tramway connecting with dedicated sidings at Corsham station.

It is probable that the naming of camps was influenced by the nearest railway stations and to minimize possible confusion with similarly named places. (Another criterion might have been to reduce the risk of mail and other deliveries going to the 'other' place.) The camp south of Swindon was at first called Draycot after the nearby hamlet of Draycot Foliat, but there were already Draycott railway stations in Somerset and Staffordshire; very soon it was known as Chisledon, a village with its own station. Longbridge Deverill was more usually named Sand Hill, perhaps for reasons of brevity, though there was a Midland station called Longbridge. In its early years the camp at Knook was named Heytesbury, which had its own station. Perhaps Bulford Station in Essex was renamed Cressing in 1911 because of confusion with Bulford Station near Amesbury. And in Sussex there was the village of Durrington with its own station, perhaps explaining why the Wiltshire camp of that name became subsumed within Lark Hill. Similarly, Corton Camp was sometimes seen administratively as part of Boyton Camp, perhaps because there was another Corton in Suffolk, again with its own station.

A 1916 emergency scheme for reinforcing troops south of the River Thames in the event of an invasion had the following entraining arrangements for troops:

* at Tidworth/Lark Hill: one train an hour from Amesbury No. 1 platform
one train an hour from Amesbury No. 2 platform
one train an hour from Tidworth
* at Codford and Sutton Veny one train an hour from Patney
one train every two hours from Wylye, Lavington, Codford and Warminster
* in the Fovant area: one train every two hours from Wylye and Codford
one train an hour from Salisbury passenger station
one train an hour from Milford goods station, Salisbury.

Among the destination stations were Epsom, Leatherhead, Guildford and Brookwood. Getting troops from Codford and Sutton Veny to Patney (14 miles as the crow flies, several more by road) would have been a challenge and it is surprising that soldiers from the Fovant area would not have entrained at Dinton or Wilton stations.

Roads and Traffic

By 1900 the technology of rail transport was well established and was to change little in the region for many years, even if the infrastructure was enlarged. In contrast, road vehicles evolved very rapidly in the first twenty years of the new century and many Wiltshire roads were transformed from chalk tracks into more or less surfaced highways busy with traffic.

Transport of supplies for the 1898 manoeuvres involved the Army hiring 2,420 civilian drivers, 4,619 horses and mules and 1,514 horses and carts, of which 1,397 men, 2,478 animals and 964 vehicles came from London; because of the Wessex harvest, only eight men, eighteen horses and eight vehicles could be hired locally!

For the 1903 manoeuvres the Army again hired civilian animals and drivers, with 5,000 'omnibus horses' forsaking their usual streets for the bleakness of the Plain. But transport problems were eased by twelve steam traction engines, each of which could haul 8–10 tons at 4mph. Less successful were three steam lorries, capable of 8mph, whose performance disappointed. The Motor Volunteer Corps (comprising civilian motoring enthusiasts) established its headquarters at the Ailesbury Arms Hotel in Marlborough, supplying forty motorcars to convey umpires as well as motorcyclists to carry messages.

All of this traffic was at the expense of road surfaces, causing headaches for the local authorities responsible for their upkeep. The following year, Amesbury and Andover Rural District Councils refused to bear any part of the cost of improvements, which eventually were paid for by the Army in the expectation that councils would then be responsible for maintenance. But in 1908 councils again expressed deep concern; before the Army's arrival on the Plain, each mile of road had cost an average of £7 10s to maintain, but in the past three years this had risen to £13 15s. The Wiltshire authorities reckoned that the 1909 manoeuvres alone caused £521 of damage; the War Office at first offered £31 towards the repairs and finally paid £150.

Gunner Ernest Snow was killed in a traffic accident in 1909. He was one of ninety-seven members of the 10th City of London Howitzer Battery who arrived by rail at Amesbury at 1.30am and set off to march to Rollestone Camp. At about 4.35, the column was in a dip in the road at Greenland Farm (half a mile south of the camp and not to be confused with Greenlands Farm 3 miles to the north-west). Dawn was breaking but remnants of a thick mist persisted. Suddenly a car from Salisbury carrying copies of the *Daily Chronicle* to the camps appeared from behind, scattering the soldiers but narrowly missing striking any. Two minutes later, just as they had reformed, a second *Daily Chronicle* car, this one from London, ploughed into them, knocking over some forty or fifty men. Two-thirds of these had minor injuries, but ten were taken to Bulford Camp Hospital, where Gunner Snow, who had suffered a fractured thigh and skull, died.

A coroner's jury returned a verdict of death by misadventure and recommended that the driver, Alfred Saytch, be censured; in the event he was sentenced to eight month's imprisonment with hard labour. The following year a *Daily Mail* driver delivering papers to Hamilton Camp was fined £10 plus £3 0s 1d costs for exceeding 20mph at Durrington crossroads; the police said his speed was 30 miles, 1 furlong and 30 yd an hour – amazing exactness! Two days after his first offence, the same man drove the same car into another vehicle at the same spot.

In September 1909 the Army put motor-vehicles to the test after some years of cautious uncertainty about them. The day after it took twelve horses five and a half hours (including an hour for feeding) to move a Territorial divisional supply train the 5 miles from Fargo Camp to Enford, the same work was carried out by three motor-lorries, each carrying 4–5 tons, in 105 minutes.

Nevertheless, the War Office continued to depend on horse power in its literal sense. A thousand of the vehicles used for the 1910 manoeuvres were horse-drawn, but with commercial firms switching to motors the Army found it increasingly difficult to hire horses from them; some of the 3,800 animals allocated to Red Force came from as far afield as Burnley and Sheffield. Hired traction-engines had to take on water far too often, every 8 miles in the case of a Wallis & Stevens model. Of nearly eighty steam tractors attached to Red Force, thirty needed repairs. But the Army's own few lorries proved satisfactory, encouraging further mechanization in 1911. Another sign of the times was increased traffic congestion caused by civilian spectators in their cars. To allow the passage of the 6th and 10th Brigades on 23 September, all roads

The Hornsby Little Caterpillar weighed 8 tons and was designed to tow a 60-pounder gun.

between the Shaftesbury–Salisbury and Heytesbury–Salisbury highways were closed to civilians.

One revolutionary vehicle was the tracked 'Little Caterpillar' made by Richard Hornsby & Sons, seen in the Amesbury area in 1910 or 1911. Its engine started on petrol and then ran on paraffin, but 'its noise and smell are abominable and very few horses will pass it'. A conversion so that the engine ran entirely on petrol left the Army uneasy about the close proximity of so inflammable a fuel to ammunition; this and doubts about its mechanical efficiency led to the machine remaining very much a prototype, though caterpillar tracks were adopted for tanks a few years later. (The Little Caterpillar is preserved at the Tank Museum, Bovington.)

In August 1914 the War Office had 507 'mechanical transport' – or motorized – vehicles, including eighty-five in Ireland and thirty-five elsewhere overseas. Southern Command, which included Salisbury Plain, had the largest share, 123, comprising thirty-eight lorries, eleven cars, two ambulances, six motorcycles, thirteen steam tractors, forty wheeled trucks, five workshop trucks and eight 'miscellaneous'. The war brought considerably more traffic to Wiltshire and an ever-growing proportion became motor-powered. Increasing damage to the county's highways became more costly and difficult to remedy, with road stone for civil use having a low priority and with labourers joining the armed services.

Traction-engines were prominent in hauling material and stores; indeed, one cannot see how otherwise these would have been moved over rough tracks to the outlying camps, especially from the north, where all approaches to the Plain were up a steep escarpment. But the engines' wheels cut up road surfaces, making them impassable to ordinary vehicles. By February 1915, 77.5 miles of Wiltshire's roads were damaged and it was estimated that it would cost £54,122 to put them 'into passable condition' and £150,000 to restore them to their original state. Local authorities felt that since the Army was doing the damage, it should carry out the repairs. Lord Kitchener took the opposite view, believing that the councils should foot the bill for patriotic reasons and because their districts benefited from the troops' presence. In its issue of 26 December 1914, *Autocar* called for 'a Kitchener of the Plain who, by making efficient roads, will do the transport work of the camps in half the time'. Its reporter had tried to drive over the roads between the camps:

> Mud or ruts – according to the temperature – are everywhere. The roads are just two trenches of wheeltrack width, about eight inches deep on the average, the bottom of the trenches being a series of ascents or descents of six or eight inches … My own advice is, do not take your own car for the trip, but hire or borrow one …

Rail, Road and Air

> The most interesting sight is the collection of transport vehicles of all kinds. Everything required at the camps – food, clothing, and material for huts – seems to come in every sort of mechanical conveyance. The highly decorated traction engine straight from the village circus now ambles along with a load of framing for huts. Motor lorries, new and old, and motor delivery vans from the London stores carry goods and provisions of all kinds, and besides the roads are carts that, having come to grief, have been unloaded and left for further consideration at some late date.

Salisbury Plain was particularly conspicuous on an Automobile Association map issued in May 1915 showing roads damaged by military traffic. From around that time the Army Service Corps transported materials for the local authority road boards, which, as well as maintaining public highways, also built and maintained roads in the camps. One such unit was the 650th Mechanical Transport Company, which formed at Bulford 1915 and was based in Wilton during the second half of the war, with detachments working from time to time at Lark Hill. At one time comprising four officers, 248 other ranks and 122 lorries (many of the last being well worn from service in

The caption to this official photograph from the Press Bureau explains how the Australian Motor Transport column is engaged in hauling stones for 'sturdy miners from South Wales' to shovel on to roads to 'meet the excessive strain of the tremendous military traffic' on Salisbury Plain.

France), the 650th worked from various bases throughout the country. Initially more than half the lorries were not fit to run, so their repair was a priority. Such was the unit's efforts that on 25 July 1917, only sixteen out of 164 lorries for which it was then responsible in Southern Command were off the road, a commendable achievement given the wear and tear to which they were continually subjected.

Also busy in Wiltshire was the Army Service Corps' 348th company, which formed in March 1915 at Wilton. At first it transported gravel and sand from pits near Romsey to Lark Hill; later it received twenty-five Napier lorries for the construction of Yatesbury airfield. Other duties included constructing artillery ranges near Lavington and Tilshead and moving goods from Amesbury Station to the ranges.

Aviation

Even more remarkable than advances in road transport was the progress in aviation, and here Wiltshire was truly to the fore. In 1899 'a complete equipment for military ballooning' arrived at Bulford to carry observers to spot the fall of shell in artillery practice and opposing forces during exercises. Two old balloons arrived at Lark Hill in 1902 for use as artillery targets. Three years later a small traction-engine 'with mechanical wind' was used on Salisbury Plain to haul down balloons and was judged to save much money compared with the cost of hiring horses. It carried a 2,000ft cable and could haul gas wagons, enabling a balloon to be refilled in the field.

Airships featured in manoeuvres but proved cumbersome and vulnerable and were destined to be replaced by aeroplanes, which became a common sight on Salisbury Plain after Lark Hill became Britain's first military airfield. In November 1908 (the month after the first officially recognized flight in Britain, by the American Samuel F. Cody), the Aero Club, which was to be granted the Royal prefix in February 1910, sought permission to 'experiment with aeroplanes' on military land. When the War Office suggested Lark Hill, the Club jibbed at paying £35 a year for a strip of ground. However, Horatio Barber rented a plot close by on Durrington Down, where in 1909 he built a tin shed to house an aeroplane of his own design. Another shed was erected for C. S. Rolls who, as well as being co-founder of the Rolls-Royce automobile company, was an aviation pioneer, but he was unfortunately killed in a flying accident before he could take possession. Later that year, George Cockburn erected another shed there for his Henry Farman plane and Captain John Fulton leased the shed intended for Rolls.

Encouraged by such private enterprise, the Army built two double aeroplane sheds on the site in 1910. It also encouraged the British and Colonial

Aircraft Company to construct a three-bay and two more double sheds on the land, with the latter carefully positioned so that they did not block the view from Stonehenge of the sun rising at the Summer Solstice. A Bristol Boxkite designed by G. H. Challenger flew from Lark Hill on 30 July 1910 and the company established its Bristol Flying School there, catering for Army officers wishing to take up military flying. They were required to pay for their own tuition, receiving from the Army £75 towards their expenses on qualification and acceptance by the War Office. To match this grant, the school reduced its tuition fee to £75 (which included insurance against breakages and injuries to third parties) in late 1911.

In September 1910 Captain Bertram Dickson, a former gunner employed by the company, used a Boxkite to observe that year's manoeuvres, either as a neutral observer for the directing staff or as a semi-official adherent of Red Force. He located Blue Force between Amesbury and Salisbury, landed to report its position to headquarters, took off, spotted Blue Force again, landed, telephoned his headquarters – and then had his plane captured by Blue Force! Next day Robert Loraine, a well-known actor and keen aviator, is reputed to have driven to Salisbury and offered to fly for Blue Force. Some reports say he flew the captured plane, others that he piloted another Boxkite or a Maurice Farman. That week Loraine is also said to have made the first wireless transmission from an aeroplane to the ground; this was close to Stonehenge, associating England's most ancient monument with modern progress. Piloting his plane with one hand and operating a Morse key with the other, he transmitted a message over a quarter of a mile to a temporary receiving station rigged up in a Bristol hangar at Lark Hill; then on 29 September he transmitted in excess of 1 mile. Morse had been first sent and received by an aircraft in the United States on 27 August that year, but Loraine's transmission was said to be variously the first over a mile and the first to a military installation.

There are several varying accounts of these exploits by Dickson and Loraine, the confusion best summed up by a *Daily Mail* reporter describing the conflicting rumours about Dickson's landing and capture: 'some said the aeroplane had been captured, some said it had not … let us each speak of that which he sees and thus shall we arrive at the truth'. Comprehensive Army reports on the 1910 manoeuvres refer only in passing to the presence of aircraft, which 'were not able to do much'. Just as dismissively, the reports claimed that 'Captain Dickson had his hands full in managing his machine' and his information 'was … mainly negative … and unimportant'. Despite breaking down near Andover, the airship *Beta* was thought to have been more helpful, providing a stable observation platform and dropping messages on to white sheets on the ground. Loraine's role with Blue Force is not mentioned in his biography by Winifred Loraine, *Robert Loraine, Soldier, Actor, Airman*.

The Air Battalion of the Royal Engineers formed in April 1911, with Number 2 (Aeroplane) Company housing its planes in the sheds on Durrington Down while the enlisted men lived at Bulford Camp and the officers stayed 6 miles away at the Bustard Inn – a distance apart that did not help discipline. In July, the facilities at Lark Hill led to it being a control point in the Circuit of Britain Air Race sponsored by the *Daily Mail*. There were thirty entrants, twenty-one starters and just four arrivals, spread over several days, at Lark Hill (and then at the eventual finish at Brooklands), the winner being 'André Beaumont' (a pseudonym for the Frenchman, Lieutenant Jean Conneau).

On 19 May 1912 Lieutenant A. E. Burchardt-Ashton's aircraft ploughed into a crowd of spectators as it landed near the Bristol sheds, killing a boy, Leonard Williams. A Royal Aero Club inquiry absolved the pilot from blame, though noting the accident might not have happened had his view not been obstructed by his radiator; it also criticized spectators for encroaching on the aerodrome in the absence of proper control.

In August the Military Aeroplane Competition at Lark Hill aimed to select an aeroplane suitable for the British Army. Private companies provided thirty-two entries, of which eight never got to Wiltshire. Those that did faced a series of tests, including one of portability by road and rail (a reflection on the limited range of aeroplanes). The winner was the flamboyant Samuel Cody in a cumbersome craft of dated design and makeshift construction. Its success, which won the pilot £5,000 in prize money, was partly due to its powerful 120hp engine, speed, range and unimpeded view. Though the judges reckoned that none of the planes was really of military value, the War Office purchased Cody's plane and several other entrants, but not two Hanriot monoplanes, whose performance some modern assessors of the tests have reckoned was the best. (One such, J. M. Bruce, describing the judges' perceptions as bordering on the 'hallucinatory', has wondered 'what on earth [the trials] had all been about', as they did not consider engine performance and potential for carrying weapons.) But the star performer at Lark Hill was a BE 2 which, as a Government design, was ineligible to take part in the trials. (Cody's plane crashed in April 1913, killing its Army pilot.)

On 12 September the War Office stated: 'There can no longer be any doubt as to the value of airships and aeroplanes in locating an enemy on land', and Lieutenant-General Sir James Grierson acknowledged that warfare would be impossible without mastery of the air. Fly-pasts by planes of the Royal Flying Corps (formed in May 1912) featured at the 1913 and 1914 Grand Reviews of troops held at Perham Down, near Ludgershall. After the first review, General Sir Horace Smith-Dorrien noted in his diary: 'First occasion on which the Flying Corps appeared in the British Army. They flew by very well.' Nine planes took part that year, and twelve in 1914.

Samuel Cody and his 'Flying Cathedral' aircraft with which he won the Military Aeroplane Competition. A mixture of two crashed aircraft, it proved impracticable for military use and itself crashed in April 1913.

The early aviators suffered many crashes and fatalities. An Italian airman, Captain Ignio Gilbert de Winckel, narrowly escaped death but broke both legs when he crashed near Fargo, to the south-west of Lark Hill, on 17 February 1912. The first two men to die on the Plain (and the RFC's first fatalities) were Captain Eustace Loraine and Staff-Sergeant Richard Wilson, who were killed on 5 July that year. Their Nieuport monoplane attempted a turn at 400ft, when its nose was seen to drop and the plane dived into the ground 2 miles east of Shrewton near a spot now known as Airmen's Corner, or, because of the memorial erected there a year later, Airmen's Cross (map reference 098429).

A month after the deaths of Loraine and Wilson, Robert Fenwick of Freshfield, Lancashire, died during the Military Aeroplane Competition when his monoplane crashed. 'All that remained of the Mersey monoplane lay in a hollow, buckled and smashed beyond recognition, and the inventor himself was dead among the ruins,' reported *The Times*. He had designed the plane himself and built it with the help of the village carpenter.

On 17 July 1913 Major Alexander Hewetson was killed near Lark Hill when he crashed in a Bristol Prier-Dickson monoplane during his test flight for his aviator's certificate after two months' training. The inquiry concluded that the accident was due to his lack of skill in banking the aircraft on a left-hand turn, which was followed by a nosedive to the ground. Hewetson was thrown out of the aircraft and killed instantaneously. A stone slab marks the spot (map reference 139439) where he crashed and there is a memorial to him on the south-eastern corner of Fargo Plantation (map reference 114426).

Probably because Lark Hill lacked living quarters and was too accessible to the public, the War Office looked elsewhere for permanent airfields, with land near Netheravon and Upavon being chosen. Flying from Lark Hill continued until May 1914, though in July a War Office memo noted that the RFC was using sheds recently purchased from the British and Colonial Aircraft Company (which the month before had vacated Lark Hill, having trained more than 400 pupils there) and that a decision was needed about their future. A few weeks later, war broke out and the sheds became part of the massive hutments erected on the site; one was promised to the Church of England Welfare Board and others accommodated men constructing hutments. Some of the sheds still stand, in Wood Road, close to a plaque commemorating Britain's first military airfield (map reference 143436).

The area east of Netheravon village was chosen for an airfield because it was clear of obstruction, with a wind on most days to lift aircraft, and there were buildings nearby at Netheravon Cavalry School that could be used during construction. The Royal Engineers' Air Battalion started work on aeroplane sheds during the winter of 1912–13. By the summer of 1913 the sheds were complete and on 14 June the RFC's Number 3 Squadron moved from Lark Hill, followed two days later by Number 4 Squadron from Farnborough. By October living quarters were ready for use and were called Choulston Barracks, the name relating to a nearby farm. Some of the huts were of unusual design for a military establishment, consisting of white panels bordered by vertical and horizontal timber bonding strips painted black.

At the end of June 1914, all the RFC's squadrons (Numbers 2 to 5, Number 1 being an airship squadron that had recently lost all its craft and most of its personnel to the Navy) concentrated at Netheravon to test the training of personnel in the air and on the ground and to examine the problems of military aeronautics in war. The programme included tactical exercises, reconnaissance in search of ground troops, photography and balloon engagement. With war barely a month away, the camp was a timely public demonstration that the RFC was a legitimate part of the Army and gave the opportunity to test its mobilization plans – which needed to be 'somewhat revised' afterwards.

At the end of 1914 Netheravon briefly became a training school for officers who had not obtained their aviator's certificates. The airfield was then used as a base for the formation of new squadrons. With the return of peace, it became responsible for disbanding operational units, before consolidating its position as a training establishment.

In April 1912 the War Office had acquired 2,400 acres of land above Upavon for £40,000 and started converting them into an air training school.

Rail, Road and Air

Attachés from the United States, Italy, France, Japan and Germany attend the concentration of aircraft and personnel at Netheravon in late June 1914, together with British officers including Brigadier-General David Henderson (third from left), who was to assume command of the RFC when war broke out five weeks later.

Costs were borne jointly by the Royal Navy (which provided many of the fitters) and the Army. On 19 June the first buildings were taken over as the Central Flying School (CFS) of the RFC. The location was chosen because of its isolation (though it was served by a good road from Pewsey Station 6 miles away) and relative freedom from sightseers, but it suffered from turbulent air conditions. The School's initial aim was to turn holders of Royal Aero Club certificates into war pilots. The first buildings were of wood weatherproofed with tar and included a two-bed hospital. High on the downs, the airfield was a spartan place to live and some staff opted to live in Upavon village, though in 1913 concrete bungalows were built for officers, who were also provided with two tennis courts and a golf course.

The School had only eight aircraft at first, four of which were airworthy, but intended to increase this number to twenty-five and to have eighty men under tuition at any one time. The first course ran from 17 August to 5 December 1912, and involved lectures, short flights and practical work in the 'sheds and shops'. Not just officers were eligible for the course; in 1913 twenty-four NCOs and men took it, fourteen passing.

Among the first staff and pupils were men destined to become revered members of the RFC and Royal Air Force, notably Major Hugh Trenchard, later 'Father of the RAF', who became station staff officer in 1912. The School appears to have quickly attained efficiency and came out well from the embarrassment in 1913 when Colonel J. E. B. Seely, the Secretary of State for War, claimed that the RFC had 120 aeroplanes in first-class order. It transpired that just forty-four were ready to fly, of which eighteen were at Upavon. In May that year the station employed eighty-three officers and 692 other ranks.

When war was declared in August 1914 the School lost many of its men and aircraft to active service before becoming an advanced training unit and the base for an experimental flight, which evaluated new equipment. As an aid to security, the Upavon–Everleigh road running through the station was sometimes closed to the public.

In his autobiography, *Wings over the Somme*, Gwilym Lewis recalled how in February 1916 he wrote home from Upavon:

> I have arrived at the one spot that every RFC man prays he shall never go to ... The thing that upsets us more than everything else is that it is nine [six, actually] miles away from a railway station, and, right out on Salisbury Plain, the most horribly cold spot on earth. However, it is the principal flying school in England and I have heard it said that it is the best in the world ... it is one of those places where the wind blows a gale, the rain comes down in sheets, and the sun shines like fury, all at the same time! ... The quarters are old wooden huts and not so bad.

Major George Merrick was the first Central Flying School pupil to die, when, coming in to land on 3 October 1913, his Short biplane plunged violently and turned on its back, throwing him out at a height of 250ft. He was buried with full military honours at Upavon Churchyard on 8 October.

Before the war, the School's pupils had suffered several fatalities, the number increasing sharply with the rush to supply pilots for the Front. One unusual tragedy befell Captain Arthur Soames, head of the experimental department, when testing a new type of bomb in Wig Wood, near Figheldean. A 100yd wire connected the bomb with a switch, but when the latter was operated, a fragment of the 500lb bomb hit Soames, inflicting terrible injuries from which he died.

The CFS was briefly renamed the Flying Instructors School in December 1919, but next year it reformed under its original name and went on to win a worldwide reputation for excellence.

A field at Manningford Bohune, 2 miles north of Upavon, was used by Central Flying School pilots to practise landings and take-offs. This was the most modest of Wiltshire airfields. Late in the war, several airfields were built in Wiltshire, mostly to train crew for bombers. Students came from all over the Empire and, after the entry of the United States into the war, many American squadrons enhanced their flying skills at them before entering combat.

Boscombe Down, a mile south-east of Amesbury, opened in 1917. Initially bell tents and two hangars were the only accommodation on the 333-acre site at Red House Farm, though there were six aeroplane sheds and a repair shed in 1919, when training at the airfield ended; the buildings were used to store aircraft until the following year, when they then housed farm machinery. The Air Ministry purchased 500 acres of land in 1925 and built a permanent aerodrome in 1928.

Lake Down Airfield, 3.5 miles south-west of Amesbury, was built in 1917–18 close to Druid's Lodge and was the furthest point on the Lark Hill Military Railway. An official photograph dated September 1918 shows large hangars and the other usual airfield buildings, with a large number of tents providing accommodation for staff and trainees. Some officers were billeted in Berwick St James. The airfield closed late in 1919 and most buildings were removed or demolished soon afterwards.

Lopcombe Corner Airfield, 8 miles north-east of Salisbury, comprised 228 acres. Known locally as Jack's Bush, it was built in 1917 to train pilots of single-seater planes. (Lopcombe Corner is in Wiltshire, though Jack's Bush and the airfield site are in Hampshire.) The technical and living quarters were on the western edge near Mount Buncas Woods. There were six hangars for aircraft and two for stores, an aircraft-repair shed and two motor-transport enclosures covering 29 acres. It closed in November 1919.

'Market Lavington landing ground' was included in a 1919 list of RAF stations and was probably used by aircraft taking part in experiments or artillery cooperation exercises.

Near Old Sarum, the War Office requisitioned land at Ford Farm in 1917, building a double line of hangars and many wooden huts for training day-bomber pilots. The airfield officially opened in August. Known at first as 'Ford Farm', it was renamed Old Sarum to avoid confusion with Ford Airfield in Sussex, which opened in March 1918. In 1921 the School of Army Co-operation moved to Old Sarum to run courses for Army officers and Royal Air Force pilots and observers.

West of Rollestone Camp and the Bustard Inn road, the RFC requisitioned an initial 50 acres of land to establish Number 1 Balloon School in 1916 to train balloon observers. At the time of the Armistice, it covered 180 acres and had 155 training staff and a student population averaging twenty-five officers and 120 men. It became the RAF School of Balloon Training in 1920. Near Tilshead, an outstation of the school was set up in 1916 and 105 Squadron had a temporary airfield east of the village. Balloons were flown from various sites in the area, with equipment being stored close to the last house on the road from Tilshead to West Down Camp. Two balloons were normally kept there and were sometimes moved through the village to the flying areas, with men desperately hanging on to the guy ropes.

Stonehenge Aerodrome, built in 1917 astonishingly close to the ancient monument and straddling the Exeter–London main road, was a training station for crews of bombers, notably the Handley Page 0/400. It also became the base of the RFC's School of Navigation and Bomb Dropping. Though most of it was a grass field, 'maintained' by a resident flock of sheep, its hangars and other buildings severely marred the Stonehenge landscape. There is a tale that the RFC wanted the monument itself demolished because it was hazardous to flying. In August 1919 the base housed the Artillery Co-operation School, which in 1920 was absorbed by the School of Army Co-operation at Old Sarum. Then the buildings were used to breed pedigree stock before being demolished in the late 1920s. Half a mile away, close to the south-west corner of Fargo Plantation, the Royal Navy Air Service Flying School taught pilots how to fly Handley Page bombers. Initially the staff comprised six naval officers and 250 ratings, with much debate about how many ship's cooks, cook's mates and cook ratings there should be. The RNAS site closed shortly after the Armistice, though its buildings remained standing for another eight years or so.

At Yatesbury an airfield opened in November 1916. Covering 260 acres, it had two almost equally sized landing grounds and was regarded as two aerodromes, West and East, with separate groups of buildings on each. The first unit to be based there, in November 1916, was Number 55 Reserve Squadron, with other reserve squadrons joining it early the next year. By mid-1917 some 150 machines were based there. The airfield closed in 1919, but reopened as a reserve training school in 1936. In 2011 original Great War hangars at West airfield were in a dilapidated state after a project to develop the site into housing collapsed financially.

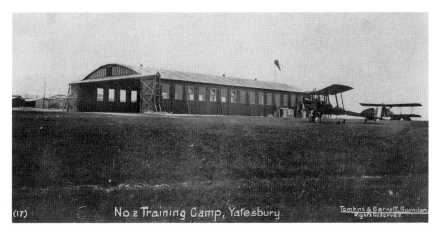

One of the hangars at Yatesbury Airfield in 1917.

Details of most of the above are given in Volume One of the *Aerodrome Board Quarterly Survey of Stations of the RAF*, published in September 1918. It lists the uses of many individual buildings, giving a fascinating glimpse of the skills need to keep warplanes in the air. At Lake Down there were huts for, among others, carpenters, sailmakers, smiths and coppersmiths. A sign of female emancipation resulting from the war was the numbers of women on the rosters. Typically, at Boscombe Down 218 of the 858 staff were female, a third being employed on 'household' duties.

Towards the end of the war an RAF stores distribution park was nearing completion on the Ham, half a mile from Westbury Station. It was intended to supply units in the RAF's South Western Area with all technical and field stores (other than complete machines and engines). In 1918 there were 282 personnel, including 125 women, on a 3.5-acre site. An 'air service camp', part of United States Services of Supply Base Section Number 3, was established at Codford on 23 September 1918, so close to the Armistice that it was presumably short-lived. An aircraft acceptance park was planned for Wroughton (on the site of the later airfield) in 1917; some preliminary work was done before the scheme was abandoned.

6
Spy Scares, Censorship and Prisons

Spy Scares

Security in Wiltshire's pre-war Army camps was almost non-existent. The main concerns related to petty crime by the troops themselves, roughs bent on trouble, hawkers, loose women and civilians taking advantage of the beer tents.

Local and national newspapers printed detailed reports of manoeuvres, which, in any case, were sometimes attended by foreign observers, nineteen countries being represented at those in 1909. The newspapers were generally eulogistic, with only minor criticisms. Soldiers in camp 'could not wish for anything' in the way of facilities, their bearing and behaviour usually 'could not be faulted', and troop movements were often 'perfectly executed'. Only the civilian horses and drivers recruited for larger-scale manoeuvres were frequently criticized, some of the former being more accustomed to hauling buses in cities than wagons on the Plain, the latter apparently unused to hard work and vulnerable to the temptations of alcohol. There were also occasional references to the lack of horses for Yeomanry and Territorials, who had to hire them for their summer camps. More revealing, as international strains grew, were references to shortfalls in manpower, such as in 1908, when it was noted that many of the new Territorial battalions were seriously under strength.

Until one was killed by a landing aircraft, civilians seem almost to have been welcome at Lark Hill Flying Ground, with photographs of the latest military aeroplanes, including one experimentally carrying a Maxim machine gun in its nose, appearing on postcards published well into 1914. But there was a hint of security awareness in the choice of the isolated downs above Upavon as the site for the Central Flying School. And it is significant that postcard scenes of the RFC concentration camp at Netheravon in the tense months of June and July 1914 show no close-up images of planes; the public were kept away and one local photographer complained when he was not allowed in the camp.

Spy Scares, Censorship and Prisons

On Britain's declaration of war, vulnerable sites such as railway tunnels and waterworks were immediately put under guard. At first, the GWR's own platelayers patrolled the line from London to Avonmouth, which ran through Wiltshire. On 15 August the military authorities took over, with National Reservists – former soldiers – from Swindon guarding lines in Wiltshire. By 9 October fourteen soldiers had been fatally injured by trains on GWR lines (despite being instructed on where they should safely stand) and two others were mistakenly shot by their colleagues.

In the first months of war there was a nationwide torrent of stories to the point of hysteria about spies, not least in Wiltshire. As early as 13 August the *Wiltshire Gazette* reported:

> Many rumours are about, including one of a startling nature as to the fate of two spies said to be caught attempting to do a diabolical act at one of our Wiltshire military camps. In the absence of official intelligence we do not give details, but the rumour is circumstantial and widely spread over the county.

A soldier at Chisledon accused of wounding a comrade in Hodson Wood pleaded that he thought his victim was a spy. A Canadian at West Down South wrote home in November that 'we caught another spy in our midst', one of a number of questionable tales emanating from the First Canadian Contingent, though it did include a number of immigrants to Canada with 'alien' origins (that is, by birth or family linked to hostile countries), and some of these were returned to Canada under escort. There were rumours of several men being arrested on the voyage from Canada, one having been found with explosives, another being shot. On the Plain a sentry was said to have challenged an intruder, who shot him in the arm. The sentry returned fire and killed the man, who was found to be carrying several vials containing cholera germs, intended apparently to contaminate the troops' drinking water. None of these stories has been substantiated.

At Tidworth Barracks 'a foreign-looking man' was arrested with a sketch described 'as the most valuable to an enemy from the aviation point of view'. At the GWR station in Swindon a man was detained after he was seen noting down the numbers and destinations of soldiers; he was able to explain himself – perhaps he was a railway official. Suspicious people were seen at Calstone reservoir, near Calne, a mile from a Marconi wireless station whose security had concerned the War Office the previous December. A member of the 7th Wiltshire Regiment at Sherrington was charged with espionage after having been seen wandering around Codford Camp in civilian clothes for several days, though he had a uniform (which, ironically, was something that many recruits there did not have).

Also at Codford, a Frome printer was arrested for taking photographs without permission in October 1914. The civil authorities discharged the man, who was allowed to keep his camera but had his film confiscated. That month at the same camp Private George Bader of the 13th Cheshire Regiment was charged with communicating with the enemy – a 'prisoner of war' (actually an interned civilian) confined in Lancashire. Bader was an English subject, the son of German parents. Such early cases of alleged spying were briefly heard by the local magistrates, who handed them back to the military to deal with, resulting in their outcomes not being revealed to the press. At Codford again, the *Daily Telegraph* of 16 November 1914 reported that two members of the Hampshire Regiment 'have been proved to be German spies and have been dealt with by the military authorities', copies of letters regarding troop movements having been found on them. One suspects these letters may have contained little more than incautious jottings by innocent parties rather than intelligence of importance to the enemy. The information in them could not have been more useful than many details given in the press in the first weeks of war.

Once the hysteria had eased a little, one or two surprising cases came to light of a more relaxed attitude by the authorities themselves. In December 1914 a *Daily Express* reporter visited the George Hotel in Amesbury. The hotel was close to Bulford Barracks and the new hutments of Lark Hill – and employed a German chef, who cooked lunch for Canadian officers based in the locality. 'I am told you have a German chef here,' the reporter said to the manager, who replied 'quite right, he cooked your lunch.' It transpired that the chef, Peter Kohler, had come to England as a boy and had three brothers fighting for England; he had been interned but released on the surety of the hotel manager, who nevertheless said he thought all Germans should be interned (unless, presumably, they were good chefs).

Rather more disturbing was the presence in Durrington of a German photographer, Frederick Rosener. Not until November 1915 was he charged with being an enemy alien and as such having four cameras without the permission of the registration officer. He had started a photographic business locally in 1913 and when applying for a trading pass to visit local camps had claimed to be a Dane. His pass was withdrawn in August 1914, but he had continued to take photographs at local camps. He admitted to the court that he was a German and said he had tried several times to enlist in the British Army. He was sentenced to six months' imprisonment with hard labour.

Almost as astonishing was the case of Otto Fritsch, who appeared at Warminster Petty Sessions in July 1917, charged with failing to register as an alien. He had been employed by the Army Canteen Committee at Heytesbury and Codford. 'I want to go to Germany to go to Kiel Harbour and join the German Navy,' he told the court. 'I know for a fact that A.C.C. is not Government but a lot of swindling.' He was a German subject, having been born in that part of Poland then dominated by Germany, and his father was interned in London. The chairman of the magistrates was amazed he should be at large in the camps.

One Wiltshire spy of the mid-war period has been named as Hacker, a school-attendance officer of Crockerton, near Warminster. Mrs G. M. McCracken, in *Looking Back on Seventy Years of Sutton Veny*, says Hacker was arrested in 1916 and shot in the Tower of London, after showing too much interest in local camps. There was such an official called William George Hacker, who lived at Crockerton Villas, but he was certainly not shot in the Tower; at most, he was probably a victim of local gossip and suspicion. (Another attendance officer, at Chitterne, 6 miles from Sutton Veny, was imprisoned, but for contempt of court relating to charges involving trust deeds.)

There were mutterings over many months about the Germanic origins of Miss Rettberg, the matron of Devizes Hospital, and these persisted as late as August 1918, when a correspondent to the *Wiltshire Gazette* came to her defence, pointing out that she had cared for a child of one of those complaining about her.

Censorship

In mid-1914 the War Office was issuing posters to be displayed within a 10-mile radius of Netheravon airfield forbidding the public from photographing aircraft at a distance of under 40yd. It commented that:

> Mr Fuller, a photographer of Amesbury, has been selling picture postcards of government aircraft taken during the concentration of the Royal Flying Corps (Military Wing) at Netheravon. From the size of the photographs, they have either been taken from within 40 yards of the aircraft or with a telephoto lens.

In fact, it had been T. L. Fuller's rival, Marcus Bennett, who had published most of the postcards featuring the concentration at Netheravon, some showing British and foreign officers, whereas Fuller produced many cards of aeroplanes at Lark Hill.

Censorship of a sort was imposed immediately on the declaration of war, with newspapers being banned from giving details of the mustering and departure of the British Expeditionary Force. On 11 August the Government set up the Press Bureau to regulate what newspapers might print. Its work mainly related to military operations overseas, but the restrictions imposed by the Defence of the Realm Act extended to local newspapers, greatly inhibiting their reports on local military activities. The Bureau maintained that censorship was 'voluntary', but penalties for editors who offended was harsh – potentially life imprisonment, though this was later reduced to six months with or without hard labour, and possible additional penalties of a £100 fine and seizure of newspaper plant. On 12 August *The Times* was still listing new Army divisions and their constituent regiments, with further details appearing on the 17th. The *Wiltshire Gazette* printed an Army Reservist's weekly observations on the progress of recruits at Codford and on 1 October named the battalions and their commanding officers based there. The same newspaper printed the admittedly bland diary of a member of the 4th Wiltshire Regiment at Durrington and West Down South camps, although its correspondent at Codford noted that 'strict press censorship forbad' it describing a drill manoeuvre that had excited the admiration of a colonel.

From the beginning of 1915, Wiltshire newspapers printed reports on local military activities with a number of blank spaces for locations, even in the case of local railway stations from which troops departed for active service. Circumspection continued for the duration, usually evidenced through careful writing rather than blanks. The outbreak of meningitis among Canadians was referred to only briefly in reports in the press by the county medical officer and the locations within Wiltshire of proposed airfields were not announced. However, news of men of the Wiltshire Regiment, sometimes with a blank for the number of their battalion, serving overseas continued to appear (being too dated to be of any use to the enemy), and reports of inquests into numerous aviation fatalities were published in detail until 1918. In July that year the *Wiltshire Gazette* announced that it was no longer allowed to disclose where flying accidents occurred. When recounting how two airmen had died when stunting for the benefit of wounded soldiers at a local country house, it dutifully referred to the latter's hosts as Sir—and Lady—of— House. However, in its next issue of 1 August it named the site of another fatal aeroplane crash as being Stert, helpfully describing the village's location in relation to Devizes. After that lapse, it maintained the required discretion.

Frustratingly for today's historians, the censorship regulations meant that after 1914 no details of the building of camps and military activities at them were printed in newspapers. But camps were named in advertisements inviting tenders for goods and services and in news reports on welfare facilities and court cases relating to soldiers and civilian workers. Sometimes press reports on incidents at camps were quaint in their discretion. These were not named when it came to accusations of waste, but if someone from Dinton complained about 'a nearby camp' it was reasonable to assume he was referring to Fovant. Again, when George V visited local camps in February 1917, they were not identified in the *Salisbury Journal*, which, however, did say he had alighted at Dinton Station and proceeded to Wilton, 'inspecting troops' along the way – which suggested Fovant and Hurdcott camps.

Prisoner-of-war camps (usually located next to, or as part of, military camps) were named when Scotland Yard released details to the press of escapees.

The resulting reports sometimes presumed local knowledge, as with *The Times* referring merely to the 'Perham Branch Camp' – a small camp at Perham Down, near Ludgershall, whose location would be known to few outside the locality.

When the 6,420-acre Amesbury estates of Sir Cosmo Antrobus were auctioned in late 1915, the three maps in the prospectus (based on pre-war surveys) were devoid of military detail, though purchasers were assured that they would benefit from any sustainable claims against the Government in respect of 'certain Railways and other works which have been carried out by or on behalf of the War Department on the Estate'.

The magazines of military units published locally were also subject to censorship. The first issue of *Direct Hit*, produced by the 58th London Division at Sutton Veny in September 1916, seems harmless enough, but the second was delayed and its original copy had to be re-edited because of censorship rules. Local towns were referred to as W—— and S—— (presumably Warminster and Salisbury). Such scrupulousness was not observed by the magazines published in 1916 and 1918 at Fovant and Hurdcott camps, which freely named Salisbury and local villages.

For a time after the Armistice, newspapers remained subject to censorship rules and those covering the Plain printed little about the unrest in many camps in early 1919. Perhaps the *Salisbury Times* had in mind these – including the burning down of buildings at Lark Hill – when it remarked: 'Late events have enlightened the Government on the mind of the soldier as the arguments of their political opponents never did'. The *Andover Advertiser* referred to trouble at Romsey and Winchester, but never uttered a word about unrest on the Plain. Nor did the national press describe soldiers' disturbances in Wiltshire, but those at Rhyl, Bristol, Dover, Ilford, Maidstone and elsewhere were reported (though these impinged more on the civilian population than did most of those at the county's military bases). The restrictions on press coverage were removed in time for details of a mutiny at Durrington in July by men of the 3rd West Yorkshire Regiment to be reported.

In October 1915 the Government pointed out to:

> all concerned in the publication, sale and distribution of picture postcards that those representing docks, harbours, shipyards, defences, ammunition works, prominent buildings, monuments and other features in or near the approaches to towns and populous districts which may afford landmarks for the guidance of enemy aircraft may be regarded as likely to assist the enemy.

Under the Defence of the Realm regulations such postcards could no longer be sold, but the huge military badges carved in the hillside near the camps at Fovant and Hurdcott were surely 'features' or 'landmarks', yet these appeared in wartime postcards, albeit uncaptioned ones. Postcards of military hutments continued to be published, but most publishers played safe by showing very general views and small groups of soldiers, captioning them (if at all) merely 'The Camp'.

Enemy Internees and Prisoners of War

When, early in the war, Westbury Rural District Council heard that 500 German prisoners of war (PoWs) were being sent to Wiltshire, a councillor suggested the downs near Imber as a suitable place for them to be held, because this area was well watered. Malice might have been behind the suggestion, because it is one of the bleakest spots in Wiltshire. Had the suggestion been accepted, disease and perhaps death would have been rife, though it could be argued that British troops endured living under canvas in the locality, albeit with a high rate of sickness.

When internment and prison camps were established in Wiltshire they were attached to military bases and airfields. A War Department list dated November 1917 mentions only two prison camps in Wiltshire, at Lark Hill and Perham Down, and an interned aliens' camp at Bulford, but a Home Office report dated January that year lists camps at Bulford, Chiseldon, Codford, Lark Hill, Fovant, Perham Down, Sutton Veny and Yatesbury. At that time, they were guarded by men of the Royal Defence Corps, who were either too old or medically unfit for active Front-Line service. In May civilian internment camps were noted at Bulford and Sutton Veny – and at Fargo and Sutton Veny hospitals, where either the inmates worked as orderlies or were patients in guarded wards.

As the war progressed and labour shortages became acute, many prisoners worked in small groups outside camp, sometimes being lodged in secure accommodation to avoid losing time in daily travel and with a small guard. They were employed on forestry work and farming, with groups on farms formed into plough teams – for example, Mr Wilson of Ramsbury had a team of thirty men. In October 1918 other work (or working) camps were recorded at Avon House, Chippenham, Netheravon, Upavon and Wootton Bassett. There was an 'agricultural camp' at Fair View, Devizes. Prisoners working on hedges and ditches at Draycot Farm had their food sent across the road from Chiseldon Camp, but with rationing this was meagre considering their heavy duties and local people would surreptitiously pass them a loaf or two. Road-making was another task for PoWs. 'They are exceedingly cheeky, and will tell you they like doing this work, for they will be using the roads themselves after the Kaiser invades England,' one Australian wrote home. In the Second World War, German prisoners working on the Kennet and Avon Canal are said to have found carved on brickwork the names and units of their Great War predecessors.

On 7 September 1917 the 'working camp' at Bulford that had opened about two months before was inspected by a representative of the Swiss 'protecting power', responsible for ensuring the inmates were being well treated. Bulford held 191 civilian internees, the majority with families in the United Kingdom, who had volunteered to work to earn money. They were housed in twenty-five Army tents, each accommodating eight men, with two marquee tents furnished with dining tables and benches; one tent served as an infirmary. The men were employed on painting, plumbing, blacksmith work, carpentry and bricklaying. Unskilled workmen earned 7d an hour, carpenters 10½d, painters 7d to 7½d and bricklayers 10½d. Weekly earnings averaged £11 6s 9d, with each man paying 2s 6d a day for his maintenance, unless wet weather prevented work. Most drew 15s a week to spend at the canteen, though the inspecting officer was told that some of this was spent on gambling. An inmate could buy a suit for 16s and second-hand boots for 4s. There were complaints that the dentist visiting from Amesbury charged £4 10s for artificial teeth. Nor did the prisoners like the herrings that apparently formed a major part of their meals. They were due to move into a hutted camp in a few weeks.

Major J. L. Isler of Switzerland visited five Wiltshire working camps in October 1917. At Codford the camp had held combatant prisoners up to 11 June, when they had been replaced by civilians, comprising 114 Germans, 24 Austrians and 19 Turks. They were employed as carpenters, plumbers and painters and in road-making, being paid from 7d to 10½d an hour and working fifty-two hours a week. At Lark Hill there were 501 'combatant prisoners', fifty of whom were sleeping in tents. On 9 September a sentry had fired into the PoW compound, wounding six men, four of whom had gone to hospital; a court of inquiry found that the sentry was mentally deranged and so not accountable. At Fovant, Sub-Lieutenant W. T. Cruickshank was in charge of 125 Army and sixteen Naval men, with five others under detention for attempted escapes. At Sutton Veny there were 200 German soldiers, who complained about the lack of a drying-room; the camp was to be closed in a fortnight's time and its inmates moved to another nearby.

At Upavon 101 German soldiers and one sailor had a *feldwebel* (sergeant) as camp captain. Their camp was 60yd square and comprised thirteen tents, eleven serving as dormitories; the *feldwebel* and camp interpreter had a tent to themselves and another was used as an isolation tent and sitting-room. The men did excavations, road repairs and fencing of roads and paths and were paid 1d, 1½ or 2d an hour (these lower rates being due to the fact that they were captured servicemen rather than civilians who had been interned as a precaution). They worked 53½ hours a week. The camp was also to be closed at the end of October, probably because the tents were unsuitable for winter accommodation.

German prisoners of war and their guards, portrayed on a card with the inscription 'In remembrance of the days spent happily together Joe Hopsdorf Swindon Mai 1918'.

A list drawn up in October 1918 of camps that might be visited to determine the work skills of prisoners notes 144 PoWs at Chisledon, 179 at Codford, 844 at Lark Hill and 194 at Perham Down.

No camp in Wiltshire met the popular conception of a PoW camp holding large concentrations of men, or experienced any fame or notoriety as did the facilities at Frith Hill in Surrey, Donington Hall or Douglas on the Isle of Man. But several did feature in the national press (though, curiously, not so much in local newspapers) when warnings were issued to the public about escaped inmates. Between May 1917 and June 1918 *The Times* listed twenty-two escapees from Wiltshire camps.

In May 1917 three Germans escaped from their camp on the eastern side of Lark Hill, near the former British and Colonial Aircraft Company sheds. They had been dressed in German uniforms, but these had been found discarded. On 13 September that year, three German soldiers and two sailors escaped from Fovant, three being quickly recaptured, the other two being caught a day or so later.

A month later, perhaps the cheekiest escape of the war, and certainly one with melodramatic aspects, was made from the camp next to Yatesbury Airfield (where in March there had been 795 inmates). Lieutenant Paul Scheumann scrambled through the barbed wire, made his way to Chippenham wearing a suit fashioned from blankets and a mackintosh bought locally and took the train to London. He went to the theatre, before registering at

Bellomo's Private Hotel in Jermyn Street as 'Thomas Hann, High-street 145, Bristol'. This aroused Signor Bellomo's suspicions because 'all Germans write the number thus', as he told *The Times*. He alerted one Sergeant Cole and the two observed the suspect guest at breakfast, when a waiter assured the hotelier, 'that man is a Bosche'. By his own account Bellomo took the initiative:

> I interrogated the lieutenant. He said, 'I am a Swiss.' I said, 'You are a German soldier: own up.' He did so quite cheerfully, and we all went to his room. Sergeant Cole asked the lieutenant to accompany him, and he smilingly agreed. I gave him some sandwiches, and said, 'You're lucky to be treated in this way. I hope you'll tell your friends how well we treat German prisoners here.' He laughed, and went off with Sergeant Cole.

In its own report of the incident, the *Wiltshire Gazette* had the policeman telling the German that his recapture 'was the fortunes of war'.

Most escapees were quickly recaptured, but the two who came closest to success were the German sailors Otto Homke and Conrad Sandhagen, who escaped from Lark Hill on 17 April 1918 and were at liberty until early May, when they were caught trying to take a boat on the south-east coast with the aim of crossing to Zeebrugge. They were dressed as civilian sailors, in blue serge clothes and high boots, and between them had an Australian shilling and nearly £1 in English silver coins. A mile from where they were caught they had hidden two bags containing biscuits, bread and other food, clothing, razor, shaving brush and knife. The men looked robust and well fed and one had a large bottle of water.

In September 1918 *The Times* reported that two Germans who had escaped from Upavon had been recaptured, followed a day later by news that two more from the same camp had given themselves up at Banbury, 60 miles away.

Six prisoners from Yatesbury appeared in court in early 1919 accused of stealing bacon fat from the factory of C. & T. Harris & Co. of Calne. They had stuffed the fat down the fronts of their trousers. The camp commandant, Captain Mursell, told the court that the prisoners had bread and coffee before leaving camp and took with them a light lunch of coffee, cheese and bread. Each man's daily allowance was 13oz of bread, 1½oz of cheese and 4oz of beef or horseflesh. This was, the captain said, 'sufficient to keep them going but it was not sufficient to satisfy their abnormal appetites. They are gross eaters.' All the accused were sentenced to two months in prison with hard labour.

Even hungrier must have been the three Germans from a 'Salisbury Plain internment camp' (perhaps at Lark Hill or Bulford) caught by Amesbury police in October 1919; they had been slipping out of confinement to kill sheep, taking the carcasses to eat back at camp.

If such anecdotes suggest that prisoners were not well looked after, there were at least two occasions when their hosts acted in ways regarded as following the best British tradition. Job Greenhill, tenant of the Cheverell Manor estate, died in 1918 when trying to rescue two inmates from the camp at Fair View, Devizes, who were asphyxiated in a whey tank. And in July 1919 when Private L. Bruckmann rescued the pilot of a British plane that crashed at an unspecified Wiltshire airfield where he was employed, he was released forthwith, had his gallantry rewarded at a ceremonial parade, presented with some money and a silver watch and given a free passage home.

Some of his compatriots were far less fortunate, 103 dying and being buried in Wiltshire, mostly from the virulent influenza of 1918–19. In January 1919 at Sutton Veny Hospital 273 of the 573 patients were German. Of graves of the Great War period there were seven at Chisledon, two at Devizes, forty-seven at Durrington, one at Fovant, thirty-eight at Sutton Veny and eight at Tidworth. (In the 1960s the bodies of German PoWs from both world wars were removed from local cemeteries and churchyards and reinterred at Cannock Chase.)

Those who survived still had a long wait before returning home. In September 1919 Hubert Jaegster wrote from tent number 27 at Lark Hill to his mother in Braunschweig:

> As you will see from this letter, I am in the same sorry state as before, so a reunion is not on the cards for the time being. I know you have been expecting me home for a long time, but you will have to be strong and patient for a few more months. It seems the present government is making no efforts whatsoever to get us back.

At the end of July there had still been 91,818 PoWs in Britain, but nearly all of these were home by the end of the year.

Prisoners of War from Wiltshire

On 18 May 1916, *The Times* published a letter from General John Hart Dunne, Honorary Colonel of the Wiltshire Regiment, reporting that some 700 parcels were being sent to PoWs from the regiment each week, a large number being packed by:

> kind ladies who act as foster-mothers to our unfortunate half-starved men and the bulk are forwarded by the wife of our colonel of our depôt [at Devizes], who has most devotedly presided over our comfort fund since the war began. She sent out 700 special gifts to our men last Christmas, which were most gratefully received.

In November 1916 a meeting was held of representatives from each district of Wiltshire to consider forming a group to undertake the welfare of the PoWs from the county imprisoned abroad – at that time about 700 men, many of whom would not have belonged to the Wiltshire Regiment. One resolution was to send them gift parcels. At first the contents were bought locally and

Spy Scares, Censorship and Prisons

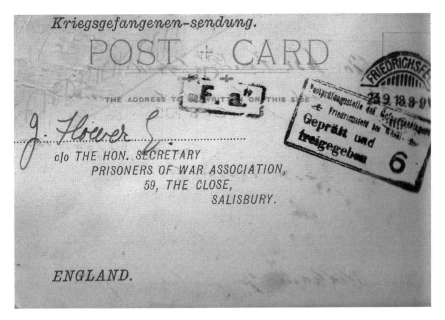

A card from Private William Taylor of the Wiltshire Regiment acknowledging the receipt of a parcel of food in good condition. 'F.a.' Stands for Frist Abgelaufen, *indicating that a period of official delay had expired (thus ensuring that any hidden messages would no longer be useful).*

then from the 'Central PoW Committee' in West Kensington and the British Red Cross Society's branch at Bristol, from where the goods arrived already packed and ready for sending. Inside an average parcel were 1lb of biscuits, ¼lb of tea, ½lb of rice, a tin each of jam, dripping, milk, Army ration, salmon and pudding, and two numbered postcards that enabled the committee to check that parcels were getting to the PoWs. The Swindon Committee started by sending parcels to about ninety men in December 1916, this number rising to 274 in October 1918.

A branch of the Prisoners of War Association was based at The Close, Salisbury (correspondence has been seen addressed to both Numbers 19 and 59), where the secretary at one time was Mrs B. E. Leech.

On 6 May 1918 Chippenham Rural District Council wrote to local parishes to raise money for the Wiltshire Regiment PoW Fund:

> Dear Sir
>
> Wilts Regiment Prisoners of War Fund
>
> With reference to the splendid work done for the above Fund last year I now enclose a Report received from the County Finance Committee for the

period ended 31 December, 1917, and your special attention is drawn to the memorandum accompanying the Report.

I am writing to ask you to again undertake the work of Collecting in your Parish, Contributions towards the Fund for sending parcels of food and necessaries to men of the Wiltshire Regiment who are Prisoners of War. The amount required last year was £17,000, and of this sum the Chippenham Rural District Council was asked to provide £1,010, and sent £1,084:7:4. The Executive Committee again appeal for not less than this amount (£1,010) and the amount asked for from the Parish of — is £ — and I ask you to be so kind as to use your best endeavours to collect this sum at your earliest convenience. The financial year will end on the 31 December, 1918, and it is hoped that the full amount, or more, may be collected and paid in before that date.

Military Criminals in Wiltshire

In the years immediately before the Great War, soldiers based on Salisbury Plain who broke military regulations served detention of up to seven days in camp, with longer periods in Aldershot and Colchester military prisons, or detention barracks such as those at Gosport and Woking, and on the Isle of Wight. Crimes against civil law could lead to sentences being served at Winchester Prison. Hutted camps had cells in which to detain minor offenders for a day or two before their cases were dealt with and, afterwards, to detain them for very short sentences. Canvas camps made do with a guarded tent.

Company commanders dealt with minor infringements of the King's Regulations, commanding officers with more serious offences. Men could be confined to barracks, lose pay, be reduced in rank or be given Field Punishment Numbers One and Two. The latter involved fatigues, pack drill and confinement to the guardroom over a period of up to twenty-eight days, with a ration of Army biscuits and water and no pay. All these penalties were included in Field Punishment Number One, which also involved spending several hours a day tied spread-eagled to a wheel. The most serious crimes were tried by general court martial, which could impose short prison sentences and the death penalty.

The increased number of soldiers in wartime Wiltshire led to Devizes Prison becoming one of a number of new military detention barracks, in its case from October 1914 to 1920. Soldiers of the 3/19th London Regiment (St Pancras Rifles) who had skipped rather too many parades at Chiseldon Camp were sentenced to fourteen days' detention there, where they had to do everything on the double with no traditional ten minutes' rest each hour on route marches and with civilian onlookers regarding them as criminals.

A further hardship was having only brown paper to smoke and just a chained Bible in the cells to read.

All Australian soldiers in England sentenced to a detention barracks were delivered to the custody of the Assistant Provost Marshal at Tidworth, who would dispatch them to whichever one had space, each man requiring an escort of at least one non-commissioned officer and two private soldiers. Some such men were suffering from venereal disease and had to be admitted to the 1st Australian Dermatological Hospital at Bulford, where security did not match that of a prison.

In 1917 there was much debate about a detention barracks specifically for Australian troops, including those suffering from VD, and needing to hold between 500 and 600 men; it was calculated that this would release more than 200 fit men from sentry and guard duties at the individual barracks. The War Office offered several prisons, including Gloucester Prison, with Devizes to take any overflow. Eventually the civilian prison at Lewes was agreed on and this was handed over, apart from twenty-four cells forming the Debtors Wing, which left 336 cells, sixty of which would be for detainees with VD who were currently held at Bulford.

The most sensational crime by a soldier in Wiltshire during the war was the murder at Sutton Veny Camp of an Australian, Corporal Joseph Durkin. Durkin shared a hut with another soldier, Verney Asser, with whom he appeared to have been competing for the affections of a young widow. On 27 November 1917 two shots were heard and Durkin's body was discovered with a bullet wound in the left cheek. Asser claimed to have been woken from sleep by the shot – though his bed was rolled up and he was fully dressed, complete with puttees and boots. He suggested that Durkin, with a fiancée back in Australia, had been depressed about his relationship with the widow and had shot himself. The death was reported to the police, whose reconstruction of the incident showed that suicide would have been impossible. (The reconstruction incorporated one of the Wiltshire police's first ballistic tests, involving firing shots into a leg and shoulder of mutton at various distances.) Asser, whose service record was more blemished than the average, was arrested and tried at Devizes Assizes, where he was convicted of murder. He was taken to Shepton Mallet Prison and hanged, after unsuccessfully applying for leave to appeal against the verdict on grounds of insanity. He was buried in the prison grounds and his very modest effects sent to his brother in Ballarat.

7
Civilians and the Army

Businesses Flourish

Few events have made more impression on the peaceful rural county of Wiltshire than the arrival of the Army. Since the Civil War 260 years before, little had happened to excite its inhabitants or to change daily life, the most evident signs of progress being the building of railways and, in the north, the associated industrialization of Swindon. Thus the arrival of tens of thousands of soldiers and their impedimenta made a major impact. Arrivals and departures of troops by train disrupted normal timetables, their movements on the roads delayed the usual civilian traffic, their camping outside a village was an attraction, distraction and nuisance all rolled into one.

The sight of columns of troops marching along behind bands must have been a rousing one, especially in the first, pre-khaki, years, when uniforms were varied and colourful. When Grand Reviews were held at Perham Down in 1913 and 1914, thousands of civilians flocked there; indeed, one reason for the reviews was to demonstrate to the populace Britain's military power. The camping season brought with it soldiers from all over the country, with the occasional visiting overseas unit providing its own individualism in an area where only Salisbury might claim any sophistication and where villagers would seldom see a visitor, or travel further than to market. During the Great War, the Australians, Canadians and New Zealanders introduced their own exotic appeal, even if this was sometimes rough and ready. Added to this, the latest technology – aeroplanes, powered vehicles, wireless equipment – was on one's very doorstep!

Spectacle apart, the Army also provided much prosperity to a county where at the end of the nineteenth century the basic industry of agriculture was in decline. Ludgershall was one of the first villages to benefit. Despite a fillip from the Midland & South Western Junction Railway having opened a station there, the community had been in decline until the Army started to camp at Windmill Hill and Perham Down and build Tidworth Barracks. In July 1898, 1,050 passengers had arrived at the station; three years later the figure was 4,900; over the same period the number of goods wagons handled there rose from 159 to 485. The *Andover Advertiser* reckoned in 1902 that the War

Office had saved Ludgershall from extinction: the station accommodation had grown and there was now a hotel and two new bank buildings. In 1901 the population was 576, in 1911 this had almost doubled to 1,117. Amesbury, Andover and Devizes also prospered from the military presence, and Tidworth and Bulford grew into small Army towns. There were many opportunities for the creation and growth of local businesses.

A good idea of one aspect of the commercial effects of massive troop concentrations is given by *The Times* of 17 October 1898, which devoted three columns to reviewing the supply and transport aspects of the recent manoeuvres involving 60,000 men in Dorset and Wiltshire. The War Office had contracted with Lipton's (a popular company with 150 stores in England and, as remained the case more than a century later, famous for its teas) to furnish three marquees at each camping-ground for every three regiments or battalions, providing porter, mineral waters, tobacco, groceries and other provisions; a bar for warrant officers and sergeants and a temperance tent were also required at each camp. Lipton's had to meet such 'multifarious needs' as bloater paste, Beecham's Pills, toilet soap and castor oil, as well as supplying 10,000lb of bread each day. Soldiers could buy ginger beer, mineral water or lemonade for 1d a bottle, a pound of biscuits for 4d, or a 4lb loaf for 6½d.

Though able to arrange all this, the company had trouble distributing provisions around the widespread network of camps. Not only were its staff inexperienced in military matters, but its 'subordinate employés were of an unreliable class', judged the Army. (On the other hand, *The Times* criticized the Army's own distribution arrangements, stating that these would have been better handled by Lipton's.) No doubt the company depended on casual labour, for it had 1,000 staff working at the various camping-grounds, with the other major contractor, Whiteley's (London's first departmental store, which was supplied by its own food-producing factories), having 1,670.

The weather in September 1898 was abnormally hot, so that troops returning to camp after a day's hard work immediately rushed to the nearest refreshment tent, irrespective of whether or not it was allocated to their particular unit. Thus some tents quickly ran out of drink. *The Times* wrote of twenty-four barmen manning a 240ft bar and trying to satisfy the needs of 3,000 customers. The temperance tents were 'practically emptied in short time' and needed a daily supply of 8,000 dozen bottles of mineral water and 150 casks each holding 36gal of ginger beer.

Local beneficiaries in subsequent years were breweries such as Simonds of Reading and Wadworth of Devizes, which expanded from selling its own beers to wholesaling a range of products needed by the troops. Arthur Yates of Ludgershall was baking 5,000 loaves a day in 1910, a quantity that rose to 20,000 during peak military activity. Hotels did well from the officer trade, while demand for postcards to send home saw photographic businesses flourish. So pleased were the trade and professional gentlemen of Pewsey with the profits generated by the Central Flying School in its first eighteen months that in 1914 they held a dinner for its permanent staff.

The barracks and camps also provided many jobs for local civilians. The railways benefited, not only from the movement of troops and supplies, but also from ticket sales to those going on leave and visiting families and friends. Tidworth became the busiest station on the MSWJR.

Not all local firms were able to respond to the larger business opportunities offered, be it catering for tens of thousands of soldiers (some suppliers being unable to gear up for a very sharp increase in trade for a few weeks each year), or building work. The construction companies for the barracks came from outside the county, though local firms gained from subcontracting and lesser projects. In the early 1900s there was, for example, the not inconsiderable work of building 300 houses in the Salisbury area for officers and others based at the 2nd Army Corps headquarters in the city.

Local people could not begin to meet the seasonal demand for horses, especially when the harvest coincided with the year's major manoeuvres. London bus companies provided many of the animals for transport and for the Yeomanry and Territorial artillery units, few of which had enough horses of their own for deployments of any size. Where the Army could not cope, the feeding requirements and other comforts of the men were usually looked after by Lipton's and other national names such as R. Dickeson & Co. and Harrods. The last offered officers 'camp equipment for annual trainings', undertaking 'the complete arrangements, providing Mess and Ante-room Tents, Draped and Decorated, Carpeted, and furnished with Settees, Easy Chairs, Writing and Card Tables, and Lighting accommodation'. The company also provided bell tents equipped with bedstead, boarded floor, chest of drawers, lantern and lantern hook, and offered a full officers' mess service under an experienced superintendent. 'Where desired, arrangements will be made for messing officers' servants.'

Wiltshire in the early twentieth century may have been a bucolic backwater, generally unaffected by national events and industrial change, but no other county had a more tumultuous Great War. To its peacetime population of 290,000 was added some 180,000 soldiers, plus several thousand refugees, foreign labourers and prisoners of war. Key factors affecting its commerce and economy were the hectic camp-building programme of 1914–15, the demand of the Army itself and so many soldiers for a wide range of services, and a shortage of workers in its major traditional industry – farming.

Though major contracts went to large national companies, local building firms such as F. Rendell and W. E. Chivers (both of Devizes, with Chivers having depots at Tidworth and Bulford in 1915) had far more business than

A fine display of food and drink for soldiers provided by Dickeson's – Army contractors who employed many ex-servicemen – at Newfoundland Farm during Army exercises in Easter 1911.

they could handle. For virtually every other type of commercial undertaking there were all sorts of opportunities. 'If there is an enterprising barber in Devizes he should come here with a dozen assistants and would soon make enough to retire on,' a Devizes Reservist on provost duties at Codford wrote in the *Wiltshire Gazette* of 24 September 1914. 'There is also a good opening for a tobacconist. Some of the small shops are taking as many pounds per day as they formerly did shillings. The proprietor of one small grocer's shop told me he had sold £50 worth of goods in one day.' One gipsy at Codford was detected ladling water out of a ditch into a dirty can, putting some powder into it and selling it to the troops as 'home-made lemonade'. Local papers also carried advertisements inviting tenders for all sorts of services – the supply of straw, hospital supplies, military funerals and a dry earth-closet service were all specified in one.

But the Reservist's plea was soon answered, with houses in Codford High Street turning over their front rooms to banks or traders. An open space in the village centre was quickly filled with shacks offering little extras to make a soldier's life more bearable, and similar shanties popped up close to most camps; at Lark Hill the shops along the Packway (the road going through the middle of the camp) were more substantial than elsewhere.

Civilians and the Army

Salisbury was an attractive centre for traders, off-duty soldiers and construction workers. Many officers from the Wylye Valley camps 14 miles and more away preferred the city to Warminster because it offered more in the way of goods and hotels. *The Times* visited the city in December 1914 and on the 22nd reported that in some cases trade had increased by 200 per cent:

> On the outbreak of war there was a breath of the general financial panic. Then the soldiers came to Salisbury Plain, and employers were glad to take back the helpers and hands they had discarded ... The shop keeper of Salisbury was in not a few cases turned into a Government contractor, chiefly for the Canadian Government ... in the market-place you may see, besides familiar farmers' carts, the row of huge military lorries. Hosiery, sweaters, blankets ... are sold in thousands to the soldiers. The hotels overflow with officers on leave, especially Canadian officers and their wives; the streets are brown with khaki; the garages choked with specialist taxi-cabs and cars.

It may surprise that wives had joined Canadians in England, but a number did find their own way across the Atlantic, causing some resentment, especially when an officer was needed urgently and he was staying out of camp. It was easier for relatives to visit British soldiers, who at Chisledon, and doubtless elsewhere, knocked on doors to see if they could be taken in. The YMCA even had a hostel at Cortington (or Corton) House for them.

Harrods provided supplies for many of the wartime camps and, shortly after war was declared, a senior member of the company eased problems at Devizes Barracks of insufficient blankets, drinking cups, plates and dishes. The company also arranged the Canadian officers' mess in 1914–15 at 5s 6d a head a day, of which the individual paid 1s, the Canadian Government the rest. The company had been told to allow for 800 officers; in fact, there were more than 1,500 and it had problems with transport, weather and staff unwilling to remain in camp. On 23 June 1915, Harrods' managing director wrote to the British Government about £6,857 owing to the company from its transactions with the Canadians, who had departed four months before. Later in the war, when soldiers of the 1st Warwickshire Regiment arrived at Amesbury at 6.30am and marched to Rolleston Camp, they found nothing to eat on arrival until 'eventually Harrods arrived with breakfast'.

Another profitable enterprise was providing hire cars and taxis. Though most of the camps were close to railway stations, civilian train services were often disrupted by military traffic. (A complaint at Sling Camp, only 2 miles from Bulford Station, was the difficulty of catching a train to anywhere.) So some garages close to camps did little else than hire out cheap cars. Hubert Andrews of the Antelope Inn, Upavon, found this a profitable sideline, and Wilfred Elkins, a retired policeman, set up a garage at Ludgershall that was run entirely for the military.

But rapid expansion proved disastrous for some. Early in 1917 William Sisley Young, a military contractor from Andover whose business interests included the Bustard Inn, found himself in receivership. Trading as G. Young & Son, he had started off providing catering at pre-war Volunteer camps. After 1914 business had boomed, giving him a monthly turnover of £50,000. But then difficulty in retaining staff and managers and changes in his financial arrangements with the Army Canteen Committee led to difficulties. His creditors finally accepted payment of 15s in the pound, with the likelihood of a further 2s 6d being forthcoming. Young valued the Bustard Inn at £6,000, though 'it could be three times that in these times'.

A number of Wiltshire's larger factories switched to war production. Cloth for uniforms was made at Palmer & Mackay's factory in Trowbridge and in Melksham Spencer & Co.'s steel foundry produced shells, while the Avon India Rubber Company turned out pneumatic and solid tyres, lifebelts, trench boots and aircraft parts. The outbreak of the war had coincided with a test involving fourteen cars equipped with Avon pneumatic tyres, which moved a battery of heavy guns at 21mph without mishap. This led to Avon products being used by all branches of the services, particularly the Admiralty, which in 1914 ordered some 40 tons of rubber sheeting and 20,000ft of hose and tubing. Avon also manufactured paravane diaphragms

Munitions workers at Spencer & Co. of Melksham, whose factory was visited by George V and Queen Mary in November 1917.

used in mine-sweeping operations. It took over the Sirdar Rubber Works in Bradford-on-Avon in 1915.

Much capacity at the massive GWR works in Swindon was switched to the war effort. The first order to the Carriage & Wagon Works was to put high sides on open wagons for conveying Army horses. There followed contracts for parts for howitzers and 60-pounder guns, locomotives, ambulance wagons, 1,100 road wagons for the Army Service Corps, water carts and aircraft bombs. In all, 238 ambulance carriages were modified or built from scratch and included staff, office and kitchen/mess coaches that were formed into trains for use at home and on the continent.

There was a large store for filled shells and shell steel at Stratton St Margaret, together with an ammunition nitrate factory whose construction started early in 1917. Labour troubles were anticipated and it was suggested that it should be a 'controlled factory' (with the suspension of any rules and customs that might limit output), but this idea was rejected, it being pointed out that 'control' would make no difference in the event of a strike, with workers still malingering if they wished. Except for a few skilled labourers, workers were recruited locally and on 11 November 1918 there were 882 men (including around 350 discharged soldiers) and 588 women, though the latter were deemed not very satisfactory or economical; they were judged not good timekeepers and stayed out for trivial reasons, although 'no doubt in a way served their purpose', as one report patronizingly put it.

Billeting

At the start of the twentieth century the Army had little experience of billeting troops on civilian households, and in the decade leading up to the war there were only a few trials, usually involving compliant civilians. But from August 1914 billeting became a necessity caused by the lack of military accommodation, especially during the wet winter of 1914–15, and soldiers were placed with householders, hoteliers and innkeepers. There were good and bad hosts and good and bad guests. In cases where the relationship was amicable it often continued through correspondence after the soldier had moved away.

But sometimes it took just one man to spoil it for the others. In late 1914 men of the Wiltshire Yeomanry billeted in a Chippenham inn were being cordially looked after by the landlady, until a late returner scoffed the meal that had been left out for her husband. At the Old George Inn twenty-seven of his comrades slept in a skittle alley, each having a mattress, bolster and two blankets.

When billeted at Mere, Gunner John Allen of the 115th Brigade, Royal Field Artillery, stabbed Gunner John Horrigan to death in a row, then tried to cut his own throat. He was acquitted of murder.

But H. W. House had every reason to be grateful to his landlady in Andover when the 7th East Lancashire Regiment moved into billets there in December 1914:

> I was lucky in being lodged with a very nice family. Mrs. Wilson was the daughter of a very well-known butcher ... She made us very comfortable, fed us marvellously and looked after me marvellously, and in fact nursed me right through pneumonia. She wouldn't allow the Regimental M.O. to move me to hospital. She said nobody as ill as I was should be moved in such very cold, wet weather. He then washed his hands of me, and I was thereafter looked after by a very good civilian doctor. I made a good recovery ... My mother was convinced that Mrs Wilson saved my life.

Soldiers in billets were usually there for several weeks or months, but occasionally the stay was for just one night, sometimes as part of an exercise. In autumn 1915 the 19th Northumberland Fusiliers were in huts at Perham Down, but on 21 and 22 October carried out 'a scheme of billeting' at Middle Wallop, Over Wallop and Nether Wallop and repeated the exercise again in late November when they took over the hamlets of Tidcombe, Oxenwood and Fosbury, close to the Hampshire and Berkshire borders. The idea was to accustom soldiers to operating as a unit when scattered around private buildings, though the main challenge was to the officers and NCOs in accounting for their men in terms of whereabouts and behaviour, as well as payments to their hosts. Earlier in 1915 the 12th Rifle Brigade had spent the night in a Middle Wallop cowshed that had an unpleasant smell; it turned out that straw had been laid 'on a lot of filth and cow dung'. The colonel sent for the farmer and his two sons and ordered them to clean out the barn and lay fresh straw within an hour, threatening legal proceedings if this was not done to his satisfaction.

Some civilians managed to make a small profit out of the billeting allowance, which at the beginning of the war was 23s 4d a week for one soldier. But on 1 September this was reduced to 2s 6d a day for the first soldier and 2s 3d for each additional man in the case of private householders. (A slightly lower allowance was paid to 'keepers of victuals', such as hoteliers.) But on 1 December 1916, rising food prices forced increases to 2s 9d for the first man and 2s 6d each for any others. The sums covered lodgings, attendance and three meals a day. If meals alone were provided (as when men were sleeping in barns or schools), the rates for breakfast, midday dinner and supper went up by 1d to 6d, 1s 2d and 4d respectively. When in autumn 1916 Dennis Wheatley, later to become the 'prince of thriller writers', was staying at the Black Horse in Tilshead, the landlady cooked very good meals and cakes for him, a subaltern, for 3s a day. (A bonus for Wheatley, a wine merchant in civilian life, was when he discovered in the cellar vintage port, unappreciated by the locals and which he was happy to buy for 2s 6d a bottle.)

In April 1917 the Army stipulated that a billeted soldier should be given a breakfast of 5oz of bread, a pint of tea with milk and sugar, and 4oz of bacon. A hot dinner should consist of 12oz of meat 'previous to being dressed', 4oz of bread and 8oz of potatoes or other vegetable. For supper, 5oz of bread, a pint of milk with tea and sugar and 2oz of cheese were prescribed.

Some accommodation was completely taken over on an extended basis. In 1914 the Army occupied many buildings in Codford and other villages nearby, using them as offices and living quarters while hutments were being built. Properties in Salisbury listed as 'military occupancies' included Eveyln House, schools at Fisherton, the Midland Bank Chambers, 3, 5, 7 and 9 York Road and 12 and 14 Wilton Road. Late in the war the Salisbury Plain training airfields were administered by the Royal Flying Corps' 33rd (Training) Wing from offices at 2A Winchester Street, Salisbury, with headquarters then being set up in Waine-a-Long Road. In late 1917 the military was occupying several buildings in the Lavington area, including the Elementary School for 'central messing', stables at Home Farm, a cottage at Cornbury Farm for use by the artillery school on Chapperton Down and the nearby chapel at 'St John-a-Gore Farm' for 'medical inspection'.

Farming

Between 1889 and 1893 Wiltshire had been much affected by a national agricultural depression: the price of wheat had fallen by 28 per cent; that of wool by 33 per cent; and income from farm rents by 47 per cent. Some tenant farmers quit, much arable land was laid down to grass and farms left to decay, so many landowners were not averse to selling their estates to the War Office. Of the thirty-two farms and twenty field barns on the central Plain between the River Avon and Tilshead some were leased to tenants, some destroyed by artillery (presumably intentionally, to provide firing practice) and others, such as Greenlands and Newfoundland farms, were used as accommodation for troops or stores.

Most farmers prospered during the war, already having benefited from an agricultural upturn in the years immediately before it, when wages and rents were stable and prices for produce steadily rising. Their main wartime problem was the loss of many of their workers to the armed services and to higher wages elsewhere, notably at the Army camps. The columns of local newspapers saw much debate about whether the needs of the country were better served by farm workers joining up or remaining on the land, a waspish view being expressed in a letter to the *Wiltshire Gazette* in December 1914: 'above [the farmers] the landed gentry have responded splendidly to the call of their country, and below them the labourers have done well, but the farmer's son simply stays at home'; the writer blamed a lack of patriotism and an unwillingness to serve in the ranks alongside labourers.

At one time, Wiltshire farmers contemplated increasing the wages only of their single men, on the grounds that their married workers depended on the farm for accommodation. In the event, the bachelors, able to earn £3 a week in the camps, laughed at a 50 per cent increase to 18s. Having made some men available to help clear up after terrible floods in January 1915, the Army thereafter hired out soldiers to farmers, or even sent them home to help with the harvest there. Wiltshire War Agricultural Committee asked for 500 men in June 1916, despite a small influx of civilians suddenly attracted to the reserved occupation of farming after the introduction of conscription. In August 1916 750 soldiers were allocated to Wiltshire farms and a year later 200 men from agricultural companies (created by the Army in February 1917 and comprising men of lower medical categories, many with no agricultural experience) were based at Devizes Barracks; in the war's last year some 1,300 troops were working on farms in the county. Farmers had to pay 4s a day for each man, half that if they provided lodgings, but some found the troops lacking in understanding of farm work and therefore of little use.

By then, however, the shortage of farm labour was even more acute because of continuing poor wages and the armed forces' insatiable demand for healthy men. A farm labourer was being paid 25s a week, so it was no wonder that men aged in their fifties were deserting the farms to earn 10d an hour cleaning roads at Chisledon Camp. Further south, a labourer could receive £5 4s 2d a week on the airfields on Salisbury Plain.

Additionally, Belgian, Portuguese and Chinese labourers worked locally. In April 1918 fifty Portuguese labourers were transferred from Worthy Down airfield near Winchester to Savernake Forest after disturbances over having to sleep in tents. Portuguese workmen also improved roads and walls at Gomeldon after an increase in traffic following the opening of the experimental ground at Porton. A court in August 1918 dismissed a case against Lam Foi, who was accused of stabbing Hop Yow in a dispute over the ownership of a bird in the Chinese quarters at Old Sarum Airfield.

The commandeering of farmland for military needs was doubly frustrating to its owners. There was its loss at a time when good money was to be made from it, and the War Office could be a difficult and bureaucratic guest. East Farm at Fovant had almost half its land – 200 acres – taken over, necessitating the sale of its flock of sheep, but the farmer, John Combes, converted some of its buildings into a camp stores and tea room, which proved so popular as to gain trade from the official canteens. He also took in officers' wives as paying guests.

Due to their absence in the Army, the bailiff of Arthur and Edwin Collins, the tenants of Manor Farm, Codford St Peter, spent much time corresponding with various Army officers about damage to gates, fences and crops. In August 1917 the long-suffering man wrote to Number 13 Training Squadron at Yatesbury claiming £15 after an aeroplane had damaged 32 acres of barley. Two months later he complained to the Australians at Codford Camp that 'very serious damage ... is being done to our Farm by troops digging trenches ... Great holes are being made on some of the best land ... it will be a very great amount of labour to fill them'. In November he feared that the rifle ranges might be extended to include the only area of roots, where 320 sheep were to be 'lambed down'. Most of the letters are firm but civil, but the bailiff's patience was obviously severely tested in his dealings with the Defence of the Realm Losses Royal Commission in London over delayed payment of rent. He claimed £984 13s 1d for almost a year's tenure of farmland by the military and eventually was awarded £377 2s 6d, plus £4 4s towards the cost of preparing the case. Meanwhile, a neighbour, Jack Stratton of East Farm, noted in his diary for December 1917: 'War office to pay £202 per an. for 107 acres of pasture and 40 acres of arable.' The farm itself was surrounded on three sides by Codford camps numbers 5, 6 and 7.

The problems of the Collins' bailiff were minor compared to those of Charles Spicer, who had owned Martin's Farm, near Grateley, since September 1915. At his bankruptcy hearing in mid-1920, he attributed his misfortunes to the 'military experiments' on nearby Porton Down, which had damaged grass and livestock, and to a War Department ban on his removing horses so that he was unable to use them for haulage, causing him a loss of £200 to £300. He had, however, been paid £30 compensation for land taken over by the Army. When the experimental station at Porton had opened in 1916, East Farm, Boscombe, was vacant, so the station's first commandant used it for accommodation and had telephone and electricity laid on from the camp.

Some farmers were unappreciative of certain camp produce: at one time, kitchen slops were killing their pigs and in some areas they did not care to fetch stable manure because of concern over the mortality rate of Army horses. (One presumes a suspicion that any disease inflicting the animals might be transmitted to the land.) However, in 1915 Henry Young of Bulford apparently had no worries, for he took 20,000 tons of manure from the camps for use on his 1,500 acres of arable.

Generally, farmers did well out of the war. At one time, A. G. Street thought that it was impossible to lose money and from his 300 acres (170 of which were arable) at Ditchampton Farm, Wilton, he was making £2,000. But in 1921 this dwindled almost to nothing. Every £1,000 he had invested in the farm in 1919 was worth £500 by 1923 and £300 four years later.

Volunteer Corps

Mindful of the threat of invasion on the outbreak of war, many communities sought to establish civilian or volunteer corps. The War Office forbade this in mid-August 1914 lest they attract men and resources away from the New Army, but three months later relented and sanctioned a Volunteer Training Corps (VTC) formed from men ineligible for the Regulars or Territorials. At first, the Corps wore uniforms different to those of the Army and was responsible for providing its own arms and ammunition. The *Wiltshire Gazette* of 3 December 1914 noted that around a hundred men enrolled at the first parade in Devizes:

> This Civilian Corps is one of the most democratic ever started in the town; the men in the ranks include doctors, lawyers, bankers, aldermen and councillors, journalists, tradesmen, artisans and lads of 16 years. All seem to have a keen desire to make themselves efficient.

The main activities were a couple of hours of drill and musketry practice each week and an occasional field day. Later, the War Office became more supportive, providing staff and subsidizing the expense of khaki uniforms. Members were used locally as auxiliaries, in particular as guards for vulnerable sites such as railway viaducts and tunnels.

With the setting up nationally of the VTC in March 1915, nine Wiltshire detachments were formed, its members being issued with a red 'GR' ('George Rex') armlet but initially no arms, ammunition or clothing. Training consisted of some rifle practice and a little close-order drill. The Corps also aimed to provide a trained body of men available for Home Service in an emergency to support Regular and Territorial troops. There were 626 men in the Swindon Volunteer Training Corps in October 1915. They drilled once a week in Park Lane and Prospect Place and also at the County Ground and provided orderlies for the Red Cross hospital, guards for Chiseldon Camp and volunteers for the Swindon Fire Brigade.

The VTC's name changed to 'Volunteer Force' in mid-1916 and the 1st Wiltshire Volunteer Regiment was formed. That year it was responsible for guarding 47 miles of railway track.

The Volunteer Act passed in December 1916 stipulated three sections for the Volunteer Regiments: A – men over military age; B – men of military age not serving in the Army (and, presumably, Navy); and C – men under military age.

Members were expected to attend fourteen drills, each lasting one hour a month until passed as efficient; failure to complete the required number was treated as being 'absent without leave'. From December 1917 the number was altered to thirty drills each quarter once a man was passed 'efficient'. In April that year three permanent instructional staff from the full-time Army had been attached to the Wiltshire unit: an adjutant, regimental sergeant-major and company sergeant-major (the last as musketry instructor). Headquarters were at Chippenham with detachments in twelve other towns, and the men were supplied with manuals and instructional stores, such as aiming-discs, sandbags and dummy grenades. The uniform was slate-grey and by July 1917 every man had a 1914-pattern service rifle. Detachments held local camps, but seldom trained collectively. Two lady clerks were appointed in May and August 1917. The 1st Volunteer Battalion provided fifty-men guards of honour at Trowbridge and Melksham when George V and Queen Mary visited the towns in November 1917.

In April 1918 a 'test mobilization' was held at Chippenham, with men being required to attend fully equipped within forty-eight hours of receiving notice. Trains were arranged in advance and a marquee erected on Westmead. This was the first time that the battalion had come together; the 797 men who paraded represented 92 per cent of the strength. Camps were held at Roundway Park, Devizes, in August 1917 and 1918 and in October of the latter year 250 officers and men gathered at Bratton Camp (the Iron-Age hillfort), the Warminster detail marching there and back, a total of fifteen miles.

With the raising in 1918 of the age limits for military service, many Section A men became eligible for service with the full-time Army. Call-up notices were issued to those liable to join the Volunteer Force, but nearly all those who received them appealed – often successfully – against them.

In May 1918 the Army Council called for men of the Volunteer Force to come forward for three months' whole-time service in Britain. On 28 June a Special Service company of thirty-nine officers and men went by train from Wiltshire to Saxmundham to guard points on the East Coast. From 1 to 3 June 1918 a course of instruction was held at Chippenham, including a demonstration on the golf links of an attack on a strong point by a platoon of NCOs, with officers acting as section commanders.

Under an Army Order of July 1918 the unit became the 1st Volunteer Battalion (Duke of Edinburgh's) Wiltshire Regiment. In November its establishment (or allocation of places) was 957, its strength 820. During its existence, 2,067 men trained with the battalion, 572 joining the Army proper.

Women's Corps

A notable contribution to the local war effort was made by the Wiltshire Women's Land Corps or Army and the Women's Emergency Corps. The former aimed to fill the gap caused by male farmhands joining the Army, By July 1916 2,656 women were employed on Wiltshire farms, with six training schools being set up for them. The agricultural community's reactions were mixed, though one farmer reckoned their tender touch made them good milkers and when the county War Agricultural Committee met in September 1917 there was endorsement of one member's remarks that 'girls were capable of doing a great deal of work which farmers originally thought they were incapable of'. Two dozen or so were based at Stanton, north-east of Swindon, cutting up trees felled by Canadian forestry battalions.

The Women's Emergency Corps was established in early August 1914 with the aim of registering voluntary women workers and directing and controlling their energies so as to safeguard the paid labour market. Early duties included collecting and distributing surplus food and helping Belgian refugees who had fled to Britain. The Government also suggested that they organize efforts to deal with 'women of notorious bad character who are infesting the neighbourhood of the various military camps', and this was certainly done in Salisbury, though to what degree of success is debatable.

8
Welfare and Women

Welfare Facilities

Regular church parades were compulsory and regarded by many soldiers as primarily an occasion when they had to pay particular attention to smartness. But few could have failed to have been moved by the assembling of almost their entire unit and the march to the place of worship, which, in the absence of actual churches, was often a pleasant outdoor area sheltered by trees, or, if the weather was poor, a large tent. One wonders how many of the hundreds of men were able to hear a sermon preached by even the most stentorian of chaplains, but the singing of hymns must have aroused emotions in all but the most soulless. Apart from the inevitable 'Onward Christian Soldiers', other popular hymns included 'Fight the Good Fight' and 'Praise to the Holiest', all accompanied by a regimental band. And there would be added poignancy if it was the last church parade before leaving England, an occasion that had many realizing the comfort of religion. Just before the 13th Rifle Brigade left the Plain in July 1915, sixty members attended Holy Communion in the Church of England marquee, compared with six the week before.

Of rather more importance to soldiers than compulsory services were the off-duty facilities supplied by religious organizations, with the Church of England being very much to the fore in providing refreshment and recreation tents during the early manoeuvres. In 1898 at Bulford it set up three large marquees, the middle one being for recreation and containing tables displaying games, magazines and newspapers, with free writing materials and a postbox. The tents on either side housed a refreshment bar and a church with seats for 300 people. Bible readings were given every evening and a service was held each Sunday.

Similar amenities became available at the new barracks at Bulford and Tidworth. A Church of England institute costing £5,000 opened at Bulford in 1901 and did sterling service for many years.* In 1913, 266,368 payments were made in ten months to its refreshment bar and in 1915 it had 1,000

---

*The word 'institute' was used for a building offering recreational or welfare facilities at an Army camp. Six per cent of the takings of such institutes were paid to the Army.

Welfare and Women

A church parade near Tidworth.

men at a time attending lectures by the chaplain on 'temperance, purity and kindred subjects'. A £7,000 Wesleyan Home opened in 1903 and ran a horse-drawn 'coffee and book car' to supply refreshments, stationery and books to the troops.

Another popular facility at Bulford was Miss Perks' Soldiers' Home, which opened in 1902, its spacious premises including a restaurant, devotional rooms, bedrooms and, available at 3d a time, bathrooms. It is sometimes assumed there was just one Miss Perks, who concentrated her activities at Bulford, but in fact there were two, Emma and Michelle, and a card sent in 1916 invited people to hear 'the Misses Perks' talking of their work in their soldiers' homes at Bulford and Winchester (this city having two) and 'on the Troopships, in the Camps', revealing even more of a philanthropic role than is usually credited. In June 1913 the home at Bulford needed £120 for outside repairs and £80 for inside work, so fundraising activities included a fête held at Prospect Hill, Nether Wallop.

At Tidworth the Wesleyan Church began to build a Soldiers' Home in 1907 at an estimated cost of £6,000; this was the first welfare facility of its type at the barracks. The Church of England opened an institute there in 1910 at a cost of £4,500, including furniture; there were refreshment tables with 120 seats, a reading and games room, three billiards tables, a concert hall able to hold 500 people, a 'Wordsworth room' for devotion (named after the

Bishop of Salisbury, the Right Reverend John Wordsworth) and quarters for the staff and chaplain. It burned down in 1917 and after rebuilding reopened in October 1918.

Garrison churches were not built until some time after both Bulford and Tidworth barracks had been completed (though nearby there were of course long-established parish churches). So at first at Tidworth the theatre was used for services, its highly polished dance floor making it challenging for worshippers to remain upright. The Presbyterian Church of St Andrew opened in January 1909, but Church of England worshippers and Roman Catholics had to wait until 1912 for their churches to be built, funds for the Catholic one of St Patrick & St George being augmented by fines levied on recruits for mistakes made during training; its foundations were laid by volunteers from the Royal Munster Fusiliers. At Bulford the delay in providing a garrison church was even longer, with an appeal for funds being launched in 1914, but then being suspended because of the war. Eventually a foundation stone was laid on 13 October 1920 with the intention that the £15,000 building, capable of seating 800 people, be a memorial to those who had trained at Bulford and who had fallen in the recent conflict.

At Lark Hill (which was merely a camping-site with very few structures until late December 1914) the garrison church, made of timber and corrugated iron, was erected by Christmas 1915.

Seemingly ubiquitous were the marquees and huts of the YMCA, which had been founded in 1844 and had become involved with the British Army in 1901 when it provided a recreation marquee at a camp at Conway. It was very popular and the idea was adopted elsewhere. Soon few summer camps on Salisbury Plain lacked a YMCA tent. During the Great War, the Association provided huts at most camps, including four in Tidworth and five at Lark Hill. On 2 September 1914, *The Times* reported:

> Splendid work is being done by the Young Men's Christian Association. Since the declaration of war they have established in spots easily accessible to the troops over 350 centres for recreation and refreshment. At each camp is at least one large marquee manned by voluntary trained workers, to which the men may resort for letter writing, reading, cheap temperance drinks, table games, sing-songs, and healthy amusements. Writing paper and envelopes are supplied free for the men's use, stamps and postal orders are for sale, and a letter-box is found in each tent. When the men draw their pay, they may 'bank' it at the tent, withdrawing it as required on production of the receipt. The workers in the tent are young men – university students in many cases – who are ready to get up in the small hours of the morning to prepare hot coffee in time for reveillé.
>
> The cost of erecting and maintaining a fully equipped tent for two months is about £500.

Welfare and Women

At Swindon, Frederick Fedarb, the area's YMCA secretary (or administrator), was able to rent for nothing from a publican a piece of ground that was normally hired out to fairground operators for five guineas a week, the only disadvantage to a teetotal body such as the Association being that it was opposite the landlord's own premises. He said he did not have the time to manage a temperance bar, but invited the Association to run it itself 'and I shall be glad if everything goes well'.

By October 1915 there were nearly fifty YMCA huts on Salisbury Plain – many financed by individuals – but said by the Association to be about half the number needed. Number One hut at Lark Hill was destroyed by fire in January 1915, with two of its staff being killed; by July the next year it had been rebuilt at a cost of £2,300, of which £560 was raised by the people of Torquay, with facilities including four billiards tables and an officers' café.

The YMCA reported that in the early months of war 'we have been able to open our hall at Sutton Veny for the use of the Argyll and Sutherland Regiment ... There is absolutely nothing for the men in this out-of-the-way place except for the YMCA and similar institutions.' A feature of this and other huts was the display of Allied flags, which were added to with great ceremony as the United States, Siam, China and Greece eventually declared war on Germany.

Inside the YMCA hut at Heytesbury. The boards announce that on Sundays there is a Bible class at 2.30 and a united service with address at 7.30 and that every evening there is a sing-song and lantern lecture at 7.30, with an evening prayer and hymn at 8.30.

One of the very few sour comments about the YMCA came from Charles Hennessey of the 2/15th London Regiment who thought the hut at Sand Hill Camp to be 'very dreary looking' and offering 'weak tea and stale wads [sandwiches]'.

With the arrival of several hundred Americans at airfields on the Plain in 1918 they were given their own YMCA facilities in marquees, which were to be replaced by portable wooden huts for winter use, but these had been installed at only two airfields by the Armistice. The American Red Cross also supplied uniforms and equipment for baseball and arranging inter-camp matches. The Americans' commanding officer at Yatesbury told the organization:

> You have filled our coffers with all the needed articles of comfort and health. You have given us an American flag to fly over our camp and a bugle to awaken our boys to the chilly blasts of Yatesbury. Please accept our thanks for these many favours and rest assured that it is our intention to call upon you freely for anything we need, knowing that we will not be denied.

The YMCA also had facilities in villages and towns frequented by off-duty soldiers. In Amesbury it purchased a café in 1917 and used Countess Farm House as a canteen from 1917 to 1920. It also had a shop and bakehouse in the town and rented the Primitive Methodist Church from 1917 to 1920. Although Appleshaw and Redenham (2 miles east of Ludgershall) were within a quarter of a mile of each other, each hosted YMCA facilities for Australian soldiers based locally. That at Appleshaw was in a rented cottage. The hut at Redenham opened in June 1917 but may not have proved viable, as it was transferred to Lake Down Airfield. The Armistice was declared before a hut at Collingbourne Kingston could be completed.

After the Women's Army Auxiliary Corps was formed in March 1917 its members were provided with their own huts by the Young Women's Christian Association and, at Bulford, by the Salvation Army; when the latter opened on 3 March 1918 it included such feminine touches as washable cream curtains, armchairs and plants on little tables.

By the end of the war the Salvation Army had fourteen huts at Wiltshire camps – though, slightly surprisingly, none in the Fovant area – offering similar facilities to those of the YMCA. One of the first to be opened was at Bulford, which provided free teas every Sunday, serving 10,000 men in one twelve-month period. Refreshments were available during the week, with more than 250gal of tea a day often being served. A noticeboard asked 'Have you written home? Note Paper is supplied free of charge.' At the Salvation Army hut at Bulford a popular hymn became known as the 'Bulford Anthem':

> When the trumpet of the Lord shall sound, and time shall be no more,
> And the morning breaks, eternal, bright and fair;

Welfare and Women

When the saved of earth shall gather over on the other shore,
And the roll is called up yonder, I'll be there.

Huts were managed by other organizations, including the Church of England, the Catholic Women's League and the Congregational Church. Many soldiers' memoirs speak gratefully of how the huts – whoever provided them – offered a refuge from Army life, giving a change from Army rations and the chances to write letters and enjoy a little recreation. But there are few references to the evangelical work of the organizations involved. A twenty-page souvenir booklet of the Australian YMCA at Sutton Veny glowingly portrays the recreational and welfare facilities there, noting only in passing that 'a Religious Work Director has been appointed'. Perhaps in their letters home and memoirs many men felt self-conscious about describing how they became more interested in religion at a time of great personal stress – away from home and family for perhaps the first time, living with strangers, being subject to harsh discipline and, as training neared its close, apprehension about active service with the possibilities of injury and death.

In late 1914 C. M. Alexander, an American evangelist, accompanied by a colleague, a pianist and two soloists, spent five days touring the Salisbury Plain camps. At West Down South Camp an audience of 1,000 Canadians was reported, each receiving a small combination Gospel and hymn book. Alexander cited 'Stonewall' Jackson (the American Civil War general) as a famed soldier who was 'one of the godliest men you ever saw', and then invited his audience to stand up for the Lord. They sang the vibrant marching chorus, 'Give your heart to Jesus', Alexander urging the men to 'sing it in this country, right into France, and on into Germany'. There followed an address about the Pocket Testament League, which Alexander and his wife had founded, with the promise of a special Testament to those who undertook to join. The signed acceptance of a Testament entailed its recipient undertaking to carry it and to read at least one chapter each day.

At Park House Camp nearly 200 men of the Welshmen there were said to have signed the League membership card within fifteen minutes, joined by ninety more by noon the next day. During the first four days of the Alexander party's visit to the Plain, it reckoned to have distributed 6,000 copies of the pocket Testament and enlisted 2,000 men in the League. The YMCA reckoned that his efforts and those of two of its local officials at Perham Down Camp in the early winter of 1914 persuaded 1,885 of 5,651 men there to 'decide for Christ'.

At Heytesbury Helen B. Woods ran her own soldiers' club. Private L. H. Davey of Codford Camp wrote in its visitors' book: 'Heytesbury is the place to be where the girls all love the boys from all over the seas and if you want a real good time go to the club where the ladies will look after you fine.' Another entry reads: 'In Remembrance of my 1st visit to Heytesbury and the Soldiers Club, where I had an enjoyable tea so nicely arranged by the ladies in charge a place which was to me on my 1st visit as a cup of water is to a thirsty person'.

If the emphasis of welfare work was on entertainment rather than religion, the Australian commandant at Sutton Veny noted that many padres devoted their time to spiritual welfare, neglecting the 'social part of their duties'. He required the 'Reverend Gentlemen to take a leading part in regard to the entertainment of the men', making his Roman Catholic padre responsible for the men's billiards room. The Venerable Archdeacon Ward visited several Australian camps in July 1917 to lecture the men about sex and venereal disease.

An early example of welfare in the war was the hospitality given to refugees from Belgium, with more than 1,000 arriving in Wiltshire from October 1914 and being accommodated in towns and villages. Lady Suffolk provided a marquee for some of them on her estate near Malmesbury. One recognized artist, Constant Permeke, stayed at Stanton St Bernard and another, Joseph Schippers (who was accompanied to England by his wife and eight children), had his own exhibition in Salisbury in 1915. Some refugees worked on constructing Army camps and one family ran a shop in Codford, where, in the camp, P. van Dyck found work with the Army as a surveyor's clerk and draughtsman.

Belgian refugees found work constructing Army camps, such as this one at Rollestone.

Civilian Hospitals

Though in 1914 the Army had its own hospitals at Tidworth and Bulford, the war necessitated the establishment of many others for the duration of hostilities as well as the use of existing civilian facilities. The British Red Cross Society had been inaugurated in July 1905 to help the sick and wounded in time of war, supplementing the care provided by the Navy and Army medical departments. During the Great War its Salisbury branch provided an increasing number of beds at several sites, receiving six soldiers with rheumatism and broken bones from Codford Camp on 11 October 1914. The camp relied on the facilities at Salisbury until it had its own hospital. From October 1914 to February 1919 the Red Cross in the city treated 5,615 patients.

The Voluntary Aid Detachment (VAD) came into being in 1909, supported by the Red Cross and Order of St John to aid the Royal Army Medical Corps in transporting and caring for the sick between hospitals and looking after them during convalescence. The first summer camps for VADs were held in Glamorgan and Wiltshire in 1911. The latter was at Tidworth Pennings, where ladies of the Winchester Division of the Red Cross were lent an ambulance and given stretchers so that they could have practical experience in First Aid. Members – men as well as women – also worked in auxiliary hospitals, a role that became particularly important in the war; within nine months of hostilities starting there were 800 VAD hospitals nationally, many in private houses.

The *Salisbury Journal* of 4 September 1915 listed VAD hospitals at Longleat, Charlton Park near Malmesbury, Corsham Town Hall, Draycot House near Sutton Benger, Swindon Baths, Salisbury's Red Cross Hospital, Salisbury Infirmary, Mere, Marlborough, Pewsey, Melksham, Ramsbury Vicarage, Trowbridge's Dorset House, Wingfield House near Trowbridge, Beltwood Dallery at Devizes and Tisbury. At various times other hospitals have been noted at Bodorgan on Ramsbury Hill, Tisbury's Trellis House, Melksham Conservative Club (which in late 1914 offered seventeen beds in the skittle alley), Heywood House near Westbury, Clouds near East Knoyle, Bemerton Lodge and Potterne's Eastwell House.

Swindon Baths had 815 patients from October 1914 to February 1915, but the glass roof made the summer heat uncomfortable and the hospital closed in July 1915.

At Mere, Grove Buildings, part of the National School, were turned into a Red Cross hospital for casualties from the Front. The first operation was conducted under the light of a cycle lamp. Later it treated tuberculosis patients. A plaque installed at the end of the war listed eighty-two staff and noted that 1,272 sick and wounded soldiers were treated at the hospital between 10 October 1914 and 28 February 1919. The commandant, Mrs M. B. White, was awarded the Order of the British Empire.

Lady Suffolk offered seventy-five beds at her home, Charlton Park near Malmesbury, but it closed in October 1915 when she sailed to join her husband serving in India. It had had 148 patients, mostly with shrapnel and similar wounds, though in early 1915 it had treated several cases of frostbite.

The Duchess of Somerset made available her house at Maiden Bradley, as did the Countess of Pembroke in the case of Wilton House, where the barn was fitted out as a theatre for short plays, with the Pembroke family, their staff and patients taking part. Some fifty-eight beds were available in 1918, the average occupancy being fifty. David Herbert, the Earl of Pembroke's young second son, was taught fishing and tree-climbing by the convalescent officers. Longleat, close to the Wiltshire–Somerset border, became a 'Military Relief Hospital'. Beds were installed in the Salon, nurses' quarters in the Bachelor Room and an operating theatre in the 'Bath' bedroom, with a library being turned into a sitting-room. The Marquess of Bath met much of the running costs. Patients produced a magazine, the *Longleat Lyre*, containing poems, stories and reports on activities.

Winsley Sanatorium, near Bradford-on-Avon, had been established before the war to look after tuberculosis patients. In 1916 it was caring for TB sufferers discharged from naval and military hospitals, who were often in too advanced a condition for treatment. Large houses at Figheldean,

Patients and staff at the hospital in Corsham Town Hall in September 1915, when it had fifty-four beds. Usually 'Hospital Blues' were worn to denote a serving soldier under hospital care and supervision, but occasionally grey uniforms were issued instead, as was the case here.

Ablington and Lavington were used briefly as Canadian hospitals in the first months of 1915.

Dr Sydney Cole of the Wiltshire County Lunatic Asylum at Devizes reported that in 1915 the war had not appeared 'to have led to any increase of lunacy in Wiltshire … there appears to have been a decrease, in spite of the apparent increase of population by influx of soldiers, labourers, etc from other parts of the kingdom to the great camps of Salisbury Plain'. However, twenty-three cases of 'lunacy' were diagnosed within the First Canadian Contingent and in 1916, six soldiers of unspecified nationality were admitted to the asylum, four with 'wounds of stress'. Dr Cole noted that 'the cases in which Mental Stress is entered as a factor include seven men reported to have suffered from war anxiety, either from the prospect of military service, or on account of sons in the army, or in other circumstances arising out of the war, and no less than 15 women who were stated to have worried about loved ones absent on military service'.

Entertainment

Entertainment was crucial to soldiers' well-being. At summer camps there was plenty of time for relaxation, with 'concerts, mock pageants, sports and games of all sorts' being organized. In August 1909 the regimental sports held at Pond Farm on the last day of camp included tent-pegging and lemon-slicing (requiring soldiers to show their dexterity with sword and lance when mounted), wrestling on horseback, and tugs-of-war. Another contest involved men saddling and mounting their horses and galloping over a course before being inspected.

Initially, few off-duty facilities were provided for those who did not care for the welfare or refreshment tents. They had to make their own entertainment – writing home, reading, or walking in the locality. The one game of chance permitted to private soldiers was housey-housey (now known as bingo), though many preferred to risk money and military discipline playing 'crown and anchor', a dice-and-board game heavily favouring the banker. The Australians were to bring with them their own game of 'Two-up', relating to how two coins landed after being spun. (It remains part of the Australian military tradition, being particularly popular on ANZAC Day.)

The Army tolerated the off-duty consumption of alcohol in moderation and allowed it to be served at summer camps. Simonds, the Reading brewery, had at least seven marquees at Perham Down Camp before the war. Soldiers were banned from public houses during manoeuvres and there was an unpleasant incident at Winterbourne Stoke in 1910 when John Benett-Stanford, who was both a Justice of the Peace and a Territorial lieutenant serving as a transport officer, tried to close the Bell Inn lest his civilian drivers succumb to

Welfare and Women

temptation. The landlord declined to do so, suggesting instead that Benett-Stanford put an 'escort' on the door to keep his men out. The officer appears to have over-reacted and 'by way of joke' – or so he later claimed – put a whip around the man's neck, dragged him outside and locked the pub. A scuffle resulted, which led to Benett-Stanford taking summons out against three men, making a total of five charges. Four of these were dismissed and the fifth adjourned *sine die*.

Benett-Stanford lived at Pyt House Tisbury (and had a local reputation for eccentricity) and perhaps had been annoyed at the relaxed attitude seen in his local inn the previous year. E. J. Springett of the 5th London Brigade, Royal Field Artillery, had captured 'a very fed up and tired trooper of the 18th Hussars', a member of the opposing force. They made their way to Tisbury where Springett thought an inn would be 'a good opportunity to water and feed my horse and my prisoner and I agreed it would be nice to have a pint ourselves although of course it was forbidden to enter a pub on duty'. Inside was a party of military policemen, but as they too 'were breaking the law we carried on and nothing was said'.

Early in the war there was much controversy about opening hours for public houses. With thousands of men with little to do in their leisure hours, village inns became swamped with thirsty soldiers, some of whom overindulged and became disorderly. At Codford in September 1914 one public house was open for four hours a day to serve thousands of men. Within weeks of mobilization, the authorities were asking that pubs in the Salisbury Plain military areas close at 9pm, a move not welcomed by the breweries. Canadian regulations forbade the sale of alcohol in their camps, but following disturbances in local villages their commanding officer, General Alderson, allowed camp canteens to sell weak alcohol for one hour at noon and three hours in the evening. The General and his staff had based themselves at the Bustard Inn, which might have caused some cynical comments from the rank and file, as well as deterring them from frequenting the premises themselves. Bruin, the bear mascot of the headquarters staff, was luckier and is said to have enjoyed beer there.

Though local people showed particular warmth to convalescent soldiers, they were not meant to serve or buy them liquor, as this might have hindered their recovery.* On 20 August 1915 the Home Office wrote to chief constables reminding them of this, and pointing out that most patients

---

*In contrast, there was widespread support for the *Weekly Dispatch*'s Tobacco Fund, which by late October 1914 had raised £10,000 towards 'smokes' for the troops, an achievement welcomed by Field Marshal Lord Roberts. For centuries there had been many warnings, mostly ignored, about the danger to health posed by smoking and in October 1916 a lieutenant-colonel in the Royal Army Medical Corps wrote to *The Times* to state that 'a large part of the functional disturbances of the heart for which men have been invalided home is due to excessive use of tobacco'.

wore hospital uniform (of blue or grey). Seven civilian workers who sold whisky to troops at Bustard Camp were each fined £3 with costs. In July 1918 two barmaids at the Bustard Inn were each fined £2 with 5s costs for serving beer to patients (probably from Fargo Military Hospital, 2 miles away).

No doubt the Army was pleased to see the counter-influences provided by the various welfare organizations offering non-alcoholic drinks. The Royal Army Temperance Association (RATA) had facilities at Jellalabad Barracks, Tidworth, comprising a refreshments room and billiards and reading room, with a plate awarded to the unit at Tidworth with the highest proportion of RATA members. In 1910 it was the 2nd Duke of Wellington's Regiment with 688 members, a remarkable 90 per cent of its strength.

In the years leading up to the war there were gradual improvements to off-duty amusements, notably when the YMCA tents hosted a range of entertainers, ranging from the more or less gifted amateur to the reasonably professional.

At Tidworth the Garrison Theatre opened in 1909 with a boxing competition and was also used for dinners, concerts, film shows and lectures. It was acquired by the Navy and Army Canteen Board in 1917, with a share of the profits going to the troops, probably as a welfare fund. Tickets were exempt from entertainment tax and in November 1917 a reminder was issued that civilians could be admitted only if they had passes showing they worked at the barracks; soldiers were allowed to bring their 'lady friends'.

In peacetime, off-duty officers appreciated Salisbury Plain rather more than the rank and file. Most had private incomes and were less affected by the limitations of being based in a rural county. Not only could they afford motorcycles or automobiles that enabled them to get away from it (and London was less than two hours away by rail), but they were more appreciative of country pursuits. Shooting and fishing facilities were excellent; the reason for an early ban on troops on exercises entering woods and coppices was to avoid disturbing game. The Tedworth Hunt was very popular and continued through the war, though the Royal Artillery Harriers, founded in 1907, had to put their hounds down in 1917, reforming a pack two years later. A polo ground at Tidworth was an early priority, being discussed in May 1898.

Wartime brought a massive increase in entertainment for soldiers, the inclusion of women among the performers being much welcomed. A typical concert included a singer or two, a comic, a few dramatic sketches and a military band playing popular tunes, ending with the enthusiastic singing of 'God Save the King'.

Parties of well-known professionals toured the camps at home and abroad, one such being that formed by Lena Ashwell, a leading British actress who raised £100,000 towards the financing of such entertainment and was awarded

The audience at a Lena Ashwell Concert at Tidworth Barracks in 1917. Lena Ashwell devoted herself to running concert parties for the troops in many parts of the world. By the Armistice she had twenty-five companies and 600 artistes.

the OBE in 1917.* Her party was just one of fifteen to entertain Australians at Number 4 Command Depot, Hurdcott, in December 1917, others including The Cooees' 'Eugene Conjuror', the Eccentrics and the Globe Trotters. The 'Allies Concert Party', which performed on Christmas Day 1917 at Number 7 Camp, Sling, comprised several soldiers and five maiden ladies from Bath who rendered as solos such popular songs as 'Roses of Picardy', 'Have you ever Loved any other Girl?' and 'Siesta Time'.

But some of the visiting entertainers were not always appreciated, as at Sutton Veny Camp where a touring trio offered a range of melodramas often on the theme of the fallen or threatened maiden, in this case played by a lady of some forty years, but well equipped by nature for her role as she appeared to be expecting an addition to her family. She also took the entrance money

---

*In 1906 Lena Ashwell had supported the formation of the Legion of Frontiersmen by her brother, Roger Pocock, describing its members as 'men for whom there is no post in the forces, they are spread all over the Empire – experienced scouts and craftsmen, who have fought in our wars and have the respect of all our great military leaders'. The Legion's band performed at the Whit Carnival in Salisbury in 1910, its members wearing hats similar to those of Canadian Mounties and three displaying Boer War medals. During the Great War, the Legion evolved into the 25th Frontiersmen Battalion, Royal Fusiliers, and served in German East Africa.

of 3d from other ranks and 1s for officers who sat in the front row – where objects thrown at the players by displeased soldiers sometimes fell. The actor playing the villain – 'Sir Jasper' – would watch her closely to check she did not pocket any coins. The hero – 'Jolly Jack the Sailor' – was seldom seen because the audience usually brought the show to an untimely end before he could rescue the 'maiden' from her fate. Nevertheless, the players continued to perform every evening except on Sundays for several weeks.

There was also much talent among the soldiers themselves. Three Canadian lieutenants put on their own melodramas in 1914, one featuring 'little Alice' (dressed in a tablecloth) and her mother in curl-papers waiting for the drunken father outside the gin-palace door. There was the poisoning act, where the wronged heroine (played by one lieutenant) caught the villain inserting Number 9 pills (famously prescribed for constipation) in her whisky and soda and, rushing to the window, saw the lights of her lover's approaching motor-car – in the form of a soapbox and with candles as lights.

The Times noted in September 1916 that 'each battalion usually has at least one man who can play the piano, and he is worked hard, while the men sit round and roar their favourite songs into the swirling clouds of tobacco smoke'. Their own concerts were sometimes ribald. For instance, in April 1916, the *Castironical*, the magazine of the 6th City of London Rifles, reported thus on entertainment at Fovant Camp:

> 'Jillo', the new revue at the Fovant Olympia, is proving an immense success. The show features several famous comedians and an entirely new set of swear words. (Vocabulary supplied with programme.) Messrs. Hadley and Stenner, contortionists and twisters, with their troupe of Derby Dandies, present a weird eccentric dance of the Salome type. Vivacious Vivien-Wood [a lance-corporal in D Company], the popular comedienne, has a part which suits her exactly, and her song, 'The Reason Why I Throw My Weight About', was well received.

Alternative attractions to soldiers at Fovant were Albany Ward's Military Theatre (one of a number run by Mr Ward at Army camps), which offered twice-nightly performances, Sundays included. Tickets were 3d, 6d, 9d and 1s. Perhaps it was here or at the Palace run by the Navy and Army Canteen Board that Donald Clarkson enjoyed a 'picture show', the first part of which included 'one girl singing very nicely' and a violinist who played music composed by Paderewski. 'The hall was crammed, must have been a couple of thousand soldiers there, hundreds of them wounded men recovering and a tremendous lot of them fellows who have been gassed.'

Several touring troupes were composed of soldiers who performed at civilian as well as camp theatres. The Kangaroos, obviously Australians, came from Hurdcott Camp and were billed as the Costume Comedy Concert

Welfare and Women

Members of the 2/2nd London Regiment dressed up to entertain their comrades at Sutton Veny Camp, where they were based from July 1916 to January 1917.

Company, comprising ten star artistes assisted by the depot band. They appeared at Salisbury's New Theatre in May 1918.

There were cinemas at many camps during the war. At Tidworth, the Electric, or Institute, adjoined the Church of England Institute and another was called the Hippodrome. A military cinema, the Palace, was built on the Packway at Lark Hill in 1915, but was burnt down by rioting troops in early 1919 and was replaced by the Garrison Theatre. There were no fewer than three cinemas serving Sutton Veny Camp. At one, in a quarry near the junction of Five Ash Lane and Deverill Road, there was a disturbance when the projector broke down and the manageress, unable to refund admission money because she had sent her husband off with it, offered free tickets. But this was unacceptable to some of the audience of soldiers who were leaving for France the next day. They set fire to the cinema and to barrels of oil under the stage. A complete cinema hall, with operating plant and toilets, was one of the first items to be sold off from Sand Hill Camp after the Armistice. It could hold 600 people. At Sutton Veny a new cinema was being built as the war ended, so must have had a very short life.

Townspeople also arranged many special facilities for the troops. At Salisbury Council Chamber in 1915, 175 volunteers offered for a nominal sum 'an exceptional tea with music thrown in' for an average of 400 soldiers every Sunday. At Swindon a soldiers' hostelry was set up at the Town Hall, providing 'ample and appetising meals' served promptly by a large staff of young ladies, and the Mayor's Camp Committee organized 104 concerts between 17 October 1917 and 18 July 1918. Many local people opened their homes to individuals or small groups of soldiers, providing them with meals and a refuge from camp life.

One local entertainment ceased in May 1915 when, after much national debate, it was decided to suspend horse racing except at Newmarket. The last meeting at Salisbury Racecourse took place at the end of the month, just before the ban. Many soldiers resented the fact that racing had continued until then, especially after a train bound for Waterloo Station carrying officers *en route* for France had apparently been delayed by a race-goers' special and they had missed their service to Dover. With so many troops based nearby, the feeling around Salisbury had been that racing should be abandoned for the duration; in any case, there were only two local meetings a year. (The national ban was later eased a little, on condition that meetings placed no demand on the railways.)

As in peacetime, sporting events were held regularly throughout the war. At Park House Camp in June 1915, Highland Games were contested by four Scottish battalions based there, with caber tossing, an officers' mule race, a piping championship and a dancing championship including the Highland Fling and Sword Dance. Football leagues were set up, sometimes by the men themselves rather than by officers, with matches between platoons and companies. Running, boxing and tugs-of-war were staple events, with football, cricket and rugby matches also popular, although when leagues for these were formed among convalescent Australians and New Zealanders the teams saw many changes in a season as members' fitness progressed and they moved from camp to camp – and back to active service. A nine-hole golf course was established at Codford.

Some units held 'gardening' competitions to brighten up the ground in front of their huts, laying out a small lawn on either side of the path leading to the door. At Sand Hill the 2/15th London Regiment 'borrowed' turves at night from grassland belonging to the Marquis of Bath, whose bailiff noticed the bare patch. He complained rudely to the 2/15th's colonel, so much so that only a reprimand was given to the would-be landscapers after they had confessed. As the war progressed and food shortages increased – and with the appearance of many hut frontages improved by successive residents – gardening efforts switched to vegetable production.

Immorality

Before the war the Army was careful to keep women out of its tented camps. In 1898 Wiltshire councillors sanctioned a military by-law 'excluding objectionable females' from the camps and in the 1900s standing orders for Salisbury Plain camping-grounds banned all women except on prescribed 'visiting days'. A Territorial colonel who brought along his own female cook for the officers' mess at Perham Down in 1913 was soon required to replace her.

Very little immorality involving soldiers was reported in pre-war newspapers, but the situation changed during the war. On 24 October 1914 Lord Kitchener had to issue an appeal to the public to help soldiers keep 'thoroughly fit and healthy'. He was explicit that they should not be 'treated' to drinks by well-wishers, but was more enigmatic when he asked the public 'to give them every assistance in resisting temptations which are often placed before them'. Rather blunter was the rector's letter published six days later in *The Times*: 'May I appeal to the womanhood of these girls and women who are haunting our camps to abandon their evil course, and to help, not hinder, our soldiers in their noble and arduous life of self-sacrifice.'

Whether the soldiers themselves felt 'hindered' is another matter, but the problem grew, with the women invariably being seen as the guilty parties. In response to these accusations, in November 1914 voluntary female patrols were organized by the Women's Emergency Corps to work with the police and military authorities in dealing with 'young girls whose flighty conduct sometimes renders them liable to quite unwarranted aspersions'. But in December 'One of Kitchener's Army' wrote angrily to the *Swindon Advertiser* complaining of girls being invited to join for 1d a week an unnamed league that instructed its members not to speak to soldiers – 'an insult to the men in the King's uniform'. At the same time, several other women's organizations, such as the Active Service League, were encouraging the shunning of young men not in uniform!

Women already of 'bad character' were quick to see the opportunities offered by camps full of young men, many away from home for the first time and increasingly conscious as the war progressed that they needed to make the most of what might be their last few months of life. But many others, 'flighty' perhaps but more or less virtuous before the war, were caught up in the excitement of the times. Women in rural wartime Wiltshire, often bereft of their own menfolk, were easily impressed by soldiers from other parts, many of whom contrasted favourably with village lads. As is remarked in A. G. Street's factual novel of the Fovant area in wartime, *The Gentlemen of the Party*, 'girls were at a premium, and many of them loved too well, only to find when trouble came that their lovers were overseas'. One of the book's characters remarks that 'half the [village] 'oomen to-day be hoors … there's one thing what 'ave got cheaper. An' that's 'oomen. They do vling it at 'ee, wi out waitin' to be asked. Dirty bitches.' Street himself wrote:

the newspapers might print articles telling how splendid the girls were, and in many cases these articles might be justified. But in any camp district during that hectic period of history, 1914 to 1918, the older men and women knew that the girls, both native-borne and war-imported, were anything but splendid.

William Lighthall of the Royal Canadian Dragoons, billeted in Shrewton in January 1915, witnessed the procuring success of a colleague, Archibald, a British immigrant with wives in Canada and England, who also ran crown-and-anchor (dice game) sessions, depriving his colleagues of any spare cash. Lighthall:

> noticed a furtive gathering around the kitchen door [of a house near the church] ... we looked through the kitchen window and there saw the reason for the crowd.
>
> Stretched out on the kitchen table, stripped bare as the day of her birth, lay a daughter of joy serving the line of eager applicants who had paid a fee to Archibald at the door. And above – the church organ played a moving hymn to lead the devout to their weekly prayer meeting.
>
> A few weeks later, a constant stream of invalids attended sick call and deeply regretted their participation in Archibald's lates[t] financial venture.

Three men of the Royal Engineers in Warminster were charged in January 1915 with an offence against Hilda Gray, who was under the age of sixteen. Sergeant-Major Frank Gilbert told the court that when sleeping at the Drill Hall he heard a scuffle and a girl asking 'How many more?' As with quite a few court cases that were adjourned there appears to have been no press coverage of subsequent hearings.

W. F. Badgley complained that in Devizes in 1917 'everywhere in the town are cackling women enticing the soldiers'. He blamed the women rather than the men, arguing the former were motivated by 'filthy lust and filthy lucre'. In reply, 'A Private' thought that 'everywhere' was an exaggeration, and Superintendent Brooks said that locally only one woman had been brought before the courts for immorality since the war began. 'What I see in Devizes frequently is the respectable mother with her son ... the faithful wife with her husband ... and the girl with her sweetheart,' he said reassuringly. But in February 1915 two women had been imprisoned for three months and one month respectively for keeping a disorderly house at Urchfont, 4 miles from Devizes. Their customers were mostly soldiers, one of whom had been seen to leave the house dressed as a woman, with his female companion wearing army uniform, this being one of a number of references in the war years to loose women affecting military clothing.

The Urchfont incident was one of the very few of its type mentioned in the newspapers early in the war and, bearing in mind Superintendent Brooks' comment, one might assume the problem then was not serious. Certainly the local press did not avoid reporting sex cases, such as indecent assaults on children, often discreetly noting that 'details of the assault were then given'. But as the war progressed the court reports reflected the relaxing of moral strictures, becoming more numerous and explicit.

In early 1918 two London girls were arrested after being caught sleeping rough in Eastleigh Wood, close to Sutton Veny Camp. They had army blankets and an Australian overcoat and cape. The chairman of magistrates at Warminster County Court intoned, 'it is quite evident to me why you came down here', but this particular press report contained no extrapolation of what was so clear to him. The girls received two months' imprisonment without hard labour.

At the same court in July 1918 Mabel Barker was charged with conveying VD to an Australian soldier and was remanded in custody for eight days to undergo a medical examination. She was later sentenced to three months in prison.

When laughter was heard in a disused hut at Number 7 Camp, Sutton Veny, four women and five soldiers were found lying on blankets. Three of the women, from Bath, had previous convictions and received a fine of £5 each or one month's imprisonment; the sentence on the fourth, with no record, was £3 or fourteen days' imprisonment. In 1919 three women were discovered in the old German internment camp at Sutton Veny, one wearing only trousers and cardigan; she had left her 'proper clothes' in the wood.

At Sarum City Sessions the Mayor had to pass judgment on a number of cases relating to indecent acts. In March 1917 a sergeant from Fovant was 'caught' on a footpath with a single young woman. At the hearing she was fined 10s; he was merely taken back to camp under escort. The same month the Mayor heard of a brothel in Downton Road, Salisbury, where Australian soldiers, many drunk, were a particular nuisance. One admitted paying 5s, another 3s, for 'misconduct', and it was alleged one woman was found 'sitting on a soldier's knee, with her boots off and her hair down'. The Bench decided that 'whatever they thought personally they must dismiss the case' because of insufficient evidence. In August a court heard how New Zealand soldiers were at risk from scabies and disease at a brothel in the Friary, Salisbury. A mother and her eighteen-year-old daughter, who was 'suffering from a certain disease', were before the court, which was concerned about the moral danger to five children aged from two to thirteen living there. The mother received concurrent sentences of three and six months, with hard labour.

Six women and one man were sent to prison in August 1918 for keeping a disorderly house at 31 George Street, Salisbury and for allowing young children and young persons to reside in a brothel. They received sentences ranging from one month to six months, after evidence was heard that a woman and man would spend from ten to fifteen minutes in a room together. The *Salisbury Times* of 9 August devoted three columns and, the next week, a long editorial to the matter. It also published a letter from the Association for Moral and Social Hygiene – which had recently contributed to the outcry over the authorities' blind eye to brothels for British soldiers in France – pointing out that the soldiers had not been punished, as had happened in Newport and Lancaster, where the courts had also named them.

This tolerance contrasted with the authorities' concern about the loss of manpower from venereal diseases. During the war more than 400,000 soldiers were treated for VD in Britain and France, enduring unsympathetic and painful treatment and having their leave and pay stopped. Men going on leave were issued with condoms, not always of the best quality, and a prophylactic Nargol, or Blue Label, kit. This was a cardboard packet containing calomel ointment (to be applied by both man and woman before intercourse) and jelly (to be applied internally afterwards after the man had urinated and washed himself, preferably using the antiseptic Condy's Fluid). This gave some protection against gonorrhoea, syphilis and chancre. But the soldier's companion might resent the time taken to apply ointment to herself, which was liable to cause her physical irritation, especially when done a number of times a day, and the man might be too drunk and over-eager to 'irrigate himself with a bucket siphon apparatus'. Returns from Hurdcott show that in March 1918 10,008 men accepted 9,916 preventatives, though in May the previous year 665 men accepted 772; one infers that some men were resolved not to give into temptation but that others thought they might indulge themselves more than once.

Lavage, or early treatment, huts were provided at most camps, with a medical orderly available at all times. Of 332 men with VD transferred from Hurdcott in the last six months of 1917, 218 reckoned to have contracted it in London, ten locally, 103 elsewhere in the United Kingdom and one didn't know. Local folklore has it that when a prostitute infected Australians they killed her and threw her body into a well at Manor Farm, Codford.

A case at Ludgershall in 1919 involved Mrs Emily Keen, Beatrice Keen and three children aged five, three and two. The *Andover Advertiser*'s report was explicit, reflecting the loosening of inhibitions about referring to sex. (Certainly before the war the paper had not much reported on immorality, though there must have been instances in an army town like Tidworth.) It referred to 'horrible and disgusting conditions prevailing in a Ludgershall bungalow and the abominable conduct of the wife of a soldier serving abroad'. The bungalow appears not to have had curtains, for Police Constable Batten was several times able to view what was going on inside. After seeing a soldier in the bedroom with Mrs Keen and the children, and another soldier with Miss Keen in the living room, he observed 'Mrs Keen and a soldier in a very intimate condition'. On another occasion a soldier was lying in bed between the two women. Men visiting the bungalow were mainly Australians, one of whom, wearing only a shirt, opened the door when Superintendent Buchanan knocked on it. When arrested, Beatrice Keen was stark naked and made no attempt to cover herself and Emily Keen was wearing only a chemise. The former was sentenced to two months' imprisonment for unlawfully assisting in the management of a brothel, the latter to two months' hard labour for unlawfully keeping a brothel and another two months' for allowing children to live there.

New Zealand politician Harry Holland claimed to have received many letters, either directly or passed on by parents and friends, from his country's soldiers based at Sling. He said that one writer alleged there were 36,000 prostitutes within 10 miles of the camp, a figure that can be strongly discounted given that the same person stated there were more than two million men in the Salisbury Plain area.

At Chisledon an early evening rail service from Swindon was known as the 'meat train' because of the prostitutes it regularly conveyed; they frequented an area of shack-shops that was known as 'Piccadilly', though this was a facetious allusion more to the emporia in London's famous locality than to the ladies of easy virtue who notoriously frequented it. In Swindon itself there were brothels in Morley Street and in buildings around Newport Street.

Other sex cases that were well reported late in the war included the attempted rapes of an officer's wife at Bishopstrow and an eighteen-year-old nurse at Fovant Military Hospital, the latter leading to an Australian private being given an eighteen-month sentence in August 1918. The same month an Army Ordnance Corps staff sergeant was acquitted of indecency after a woman cyclist claimed he had appeared stark naked in front of her near Great Durnford. She had 'had a good look at him' and later picked him out at an identity parade, but the case was dismissed because of insufficient evidence.

There were happy outcomes to wartime relationships, with some hundreds of English women sailing to Australia (and other countries) as brides. But many others were left behind to account to husbands, sweethearts and brothers who had returned from the war only to hear gossip about their womenfolk's behaviour and perhaps to find a baby whose father was now a long way from Wiltshire.

9
Postcards and Postmarks

Postcards

Picture postcards are a major source of illustrations of early military activity on Salisbury Plain and the most convenient to reproduce today, though often better photographs of major exercises appeared in magazines such as the *Graphic*. Other relevant illustrations appeared in *Punch*, in the form of cartoons of summer camps by Leonard Raven-Hill, who lived in Bromham from about 1896 to 1912 and was an officer in the 2nd Volunteer Battalion, Wiltshire Regiment.

The War Office's acquisition of land in Wiltshire in 1897 came very shortly after picture postcards became popular and was very much to the benefit of local photographers. On their doorsteps each year were thousands of men a long way from home, often not adept at writing letters but anxious to communicate with their families. With the telephone not yet readily available and certainly not in the wilder parts of the Plain, the preferred way was for a husband, brother or sweetheart to mail home a postcard of the camp where he was based, perhaps featuring the sender himself. Photographers would often turn up on the first day of a summer camp, snap a few posed soldiers and general shots of the tents and return the next day with a supply of cards. They were also quickly on the scene of a news incident, such as to feature the remains of a crashed aeroplane, a military funeral and even the carcass of the Royal Garrison Artillery horse struck by lightning at Rollestone Camp during the great storm of 6 June 1910! Through their efforts, such men have left behind a marvellous pictorial record of life in Wiltshire in the years before the Great War, the historical value often being augmented by the messages written on the backs of the cards.

Most of the photographers worked very close to home, many being reliant on bicycles as transport before the advent of the automobile. Cycles of the early twentieth century were sturdy machines, and would have needed to be in order to carry the heavy camera, tripod and glass plates over the rough tracks of the Plain.

Postcards and Postmarks

The best known of the local men was Thomas Lionel Fuller, who came to Amesbury in 1911 and after just two weeks as assistant to Marcus Bennett, the Bulford Camp photographer, set up his own business, recording many activities in and around the town. After serving with the Royal Flying Corps during the war, he returned to Amesbury and commercial photography.

Fuller – known as John – rarely disappointed with his photographs. He was well placed to take interesting pictures – of military stores being unloaded at his local railway station, the latest aircraft at Lark Hill and the Hornsby 'Little Caterpillar' tractor when it visited the Amesbury area, for example – but even when photographing a humdrum camp scene he often outshone his rivals. His prints were usually sharper, too, with some others' efforts being amateurish in terms of composition and production. Sir John Jackson commissioned him to photograph his firm's building of camps near Amesbury.

Other photographers and card publishers included Albert Marett of Shrewton; Tomkins & Barrett of Swindon; Talbot's Army Stores of Codford, whose photographer was on hand in 1915 to record the flooding of the stores themselves; and Fielder and Vowles, both of Warminster. The latter is said to have ceased trading before he could deliver some photographs that had been paid for by their soldier subjects; the divisional provost marshal confiscated those that had been developed so that they could be claimed by the men shown in them. Augustus David Bailey regularly visited Fovant, where he

The photographic premises of Marcus Bennett at Bulford Camp. The writing on the door announces that the hut is encroachment 702, an encroachment being a building used by civilians at an Army camp.

hired a room in which to take portraits of soldiers. He also photographed the military badges carved in the nearby chalk hillside, apparently adding his name immediately underneath that of the Australian Imperial Force; certainly 'Bailey' can be discerned there on some postcards. Another local photographer was Fred Futcher of Fisherton Street, Salisbury, who in early 1918 was receiving a petrol ration of 'not more than two gallons a month' to ply his trade.

Certain publishers operated nationwide and were well-known. Gale & Polden, for example, who in 1903 produced an early set of cards depicting Army life in Wiltshire, were based at Aldershot and their various series featuring the badges, history and traditions of Britain's regiments were very popular. Another major company whose cards include those of the military in Wiltshire was Kingsway (a brand name of W. H. Smith & Son).

During the first months of war a great many cards were published showing and naming the new hutted camps, though very soon, due to Government regulations, many were entirely uncaptioned or merely stated that the scene showed 'The Camp' (which makes identification a challenge for today's collectors). Security considerations were offset to some extent by the local photographer sometimes including his own name and town, postmarks naming the camp or the sender adding a 'giveaway' message and his address. But such 'clues' can mislead: some cards of Chiseldon Camp were published by J. H. Simpson of Andover, 28 miles away, for example, and a proportion of correspondents posted cards illustrating their last or neighbouring camps, rather than the one at which they were now based.

The prohibitions on photography appear to have been variously interpreted. There is a dearth of cards featuring some camps (notably those in the Fovant area) from early 1915 and many of those that were published feature photographs apparently taken from public roads, such as the Packway running through the Lark Hill hutments. Interestingly, given the prosecution of a man who took photographs at Codford in the first weeks of the war, that camp was often to be the subject of postcards for the next four years. Astonishingly, the Swindon firm of Tomkins & Barrett in 1917 was allowed to publish at least sixteen postcards of buildings and aircraft at Yatesbury Airfield; in contrast, there appear to be no wartime postcard photographs of the Salisbury Plain airfields.

The war saw one or two complaints about the vulgarity of some postcards, which was seen as further evidence of falling moral standards. A few did feature very mild forms of the sort of humour that would be depicted in later years on seaside postcards, but these had been available before the war. A popular theme was that of soldiers canoodling with girls. One card, postmarked 1913, shows a shapely young lady in a tight dress about to stoop to pick up a handkerchief and warns 'You have to be careful at Bulford Camp.'

Several welfare organizations produced their own cards of local camps – the YMCA, for example, and the New Zealand War Contingent Association, which published under the Aotearoa brand name. The profits from sales went towards facilities for soldiers; some cards had a penny added to their prices as a form of 'welfare surcharge'; others were provided free.

One disappointment is that though many contemporary local postcard publishers featured early aviation, they did not do similar justice to road and rail vehicles. There are very few cards featuring military lorries on the Plain, though some published in wartime show long lines of them drawn up in the centre of Marlborough High Street. A few cars used by senior officers do feature, and traction engines were a popular subject.

Even more frustrating is the dearth of cards showing wartime military railway lines with locomotives, coaches and wagons. Before the war, photographers were often at the station at Ludgershall – and, to a lesser extent, at Amesbury Station – to record the arrival of troops, with railway coaches visible in the background. The photography of military trains was forbidden during the war and only a very few wartime cards show locomotives in a military setting, such as one of the King's train at Ludgershall during a visit to troops in the neighbourhood. Soldiers who purchased these at the time for 1d or 2d would be amazed that a century later some collectors were willing to pay up to £70 for such items, though others have to be content with spending only a few pounds on cards showing unoccupied track running through Tidworth Barracks or Lark Hill Camp, for example.

The heyday of picture postcards covering every scene imaginable ended with the war, after which their emphasis was on pleasant villages, seaside resorts and beauty spots, rather than military activities. Perhaps, too, the authorities disapproved of civilians photographing soldiers and exercises, but, whatever the reason, the days of the local photographer visiting the local barracks or summer camp and taking a few customer-specific pictures all but vanished.

Postmarks

With many thousands of soldiers participating in manoeuvres in locations often remote from towns and villages, the Army set up its own post offices to deal with their mail, working with the civilian postal service to determine what local facilities were necessary on a year-by-year basis. Usually the General Post Office would supply its own trained clerks, with the Army supplying accommodation and orderlies. Field Post Offices (FPOs) were based at summer camping-grounds for several months, applying to outgoing mail postmarks bearing the camp name. During manoeuvres Army Post Offices (APOs) were mobile and followed concentrations of troops around as they moved from place to place. APOs had their own numbered postmarks that did not name the place of posting, though specialist reference works record (up to a point) where and when particular numbers were used.

In 1906 an FPO at Perham Down dealt with 16,227 items posted by 3,500 men from 21 to 28 July and with 44,321 letters, cards and parcels sent by 8,000 men from 4 to 13 August. For the 1910 manoeuvres sixteen APOs were established under the direction of P. Warren of the General Post Office, who was aided by two officers and seventy NCOs and men. They made two or three daily deliveries to troops, who posted 1,346,000 letters and 10,000 parcels and bought postal orders and stamps to the value of £3,208.

The first military postmark issued for use during manoeuvres in Wiltshire is said to be that of the 'Pewsey Field Force' in 1872, though some query if the mark ever existed. Bulford Camp had its own from 1898, and other early postmarks include Park House (1900), Perham Down (1900) and West Down (1899). Sought-after camp postmarks include those for Porton Experimental Station, Porton Experimental Ground, the Central Flying School at Upavon and Bowood Park (where Yeomanry troops camped in the summer of 1915).

During the Great War every major Wiltshire camp had its own military post office and, with the surprising exception of Codford, its own postmark. There, outgoing letters and most parcels received the civilian Codford St Mary date stamp (though some parcels did receive a simple circular mark bearing the camp name). In September 1914 Codford post office was extended and an extra pillar box installed, though this could hardly have been sufficient to handle the mail of some 15,000 recruits camped locally. In contrast, the far smaller camp nearby at Boyton had its own postmark, as did that at Sherrington, which existed for only three months.

Outgoing mail of Australian Imperial Force troops in Wiltshire received numbered AIF postmarks. Several reference works seek to identify which camp is indicated by a particular number and when it was used. As with attributions for the British pre-war Army postmarks used on manoeuvres, anomalies and omissions are not uncommon. At first, incoming mail for Wiltshire camps was sent by the Australian Base Post Office in London to Tidworth for distribution within the county. But the poor rail service soon led to it being switched to Salisbury, where the civilian post office forwarded it to local camps until January 1917, when an Australian post office was opened in the city. Each week in 1916–17 on average some 43,200 letters, 660 registered articles and 935 parcels were arriving in England for Australians, who were sending out 30,000 letters, 230 registered articles and 369 parcels. One problem was that convalescent soldiers often moved from camp to camp within Wiltshire as their fitness improved and then went overseas. Though the Base Post Office maintained a card index, updating it took a while. One Australian

Postcards and Postmarks

These four military postmarks include one used during the extensive manoeuvres of 1910, when a mobile Army post office handled troops' mail, and another used at the Central Flying School, Upavon.

at Fovant finally received a parcel that had been posted from home six months before and had criss-crossed the Channel 'three or four times'.

An efficient mail service to and from camp was vital to soldiers' morale. Donald Clarkson told his wife:

> You could never believe the excitement in [Fovant] camp when a mail from Australia comes in. There is much more excitement than on pay day or news of anything doing on the Western Front, and the fellows who don't get any have faces a foot long.
>
> All of us over here have just a longing to hear news from home and you need never be afraid that your letters are uninteresting.

One can imagine the problems of ensuring that letters and parcels, many from overseas, reached men who might be spending only a few weeks at a particular camp before moving elsewhere, perhaps on active service. For his efforts, A. H. Trinder, who was in charge of Chisledon Camp Post Office for two years, received the Order of the British Empire in 1919.

10
Prose and Poetry

Factual Works

Countless factual works mention the Great War in Wiltshire, notably regimental histories, many of which devote a chapter or so to training in Wiltshire – though others disappoint with just a passing reference on the lines of 'after hard training on Salisbury Plain, we left for France'. Official and other histories of the Canadian, Australian and New Zealand armies also give much space to training and convalescence at camps on the Plain, with a gratifying number of books written by or about members of the First Canadian Contingent. Experiences at Wiltshire camps are also described in personal memoirs, notably *A Very Man: Donald Clarkson 1880–1918* by Gresley Clarkson; *Khaki and Rifle Green* by Henry Prittie (Baron Dunalley); *Echo of the Guns* by Harry Siepmann; *Officer and Temporary Gentleman 1914–1919* by Dennis Wheatley; and *The Bottom of the Barrel* by F. A. J. Taylor.

In 1983 N. D. G. James published *Gunners at Larkhill*, followed by *Plain Soldiering* in 1987, which remain definitive works. The present author fully acknowledges James' mass of detail and his excellent bibliographical references, which have greatly aided his own research. Books have also been written on Chiseldon and Codford camps, Tidworth Barracks, the Central Flying School at Upavon and the Porton experimental establishment, in each case covering both world wars, leaving of the major establishments only Bulford Barracks needing a detailed history of its existence for more than a century.

Novels

Fovant was the basis of the fictional village of Sutton Evias, described in the novel *The Gentlemen of the Party* by A. G. Street, who farmed at Wilton in the first part of the century. He describes the arrival of camp contractors in November 1914, hut building, camp labourers' wages greatly exceeding those of farm workers, the effect of 18,000 troops living in the neighbourhood, a decline in morals and unruly Australians. The novel appears to be an authentic account of what actually happened, though it states that the military badges were cut in the hillside in 1919 to keep the troops occupied, when in fact most were created during the war. One wonders how far identifiable people were the basis for the characters, such as the farmer's daughter who charged for her favours.

In John Masters' novel *The Ravi Lancers*, Major Warren Bateman of the '44th Bengal Lancers' takes an Indian officer to his home at 'Hangerton-cum-Shrewford Pennel' in the Vale of Pewsey. They get off a train at Woodborough Station. Walks on to Salisbury Plain are described and there are very brief passing references to the Central Flying Station, Upavon, and recruiting at Devizes. In Masters' *Man of War*, the hero, Bill Miller, comes from 'Pennel Crecy' (perhaps Pewsey), 22 miles across the Plain from Salisbury. Having won a wartime commission in the 'Queen's Own Wessex Rifles', Miller returns to the regimental depot in Salisbury in June 1919 and decides to make the Army his career. A few pages describe peacetime soldiering at the depot.

Kim Kinrade's novel, *The Salient*, refers to Canadian troops arriving at Patney & Chirton Station and marching along 'old Roman cobblestones' to 'Salisbury training camp, just west of London'. It is unlikely any road they may have taken had Roman origins and any cobblestones would have long gone.

E. V. Thompson's novel, *The Lost Years*, describes the flying and love exploits of Perys Tremayne, a young pilot who trains at the Central Flying School, Upavon. Even before he passes out from the School, he has won a Russian medal, a Royal Humane Society Bronze Medal and the Military Cross! Such is Tremayne's flying skill that he is chosen to return to the CFS with a new Spad aircraft for assessment.

John A. Lee's *Civilian into Soldier* is a novel, obviously fact-based, about the war experiences of a New Zealand soldier, John Guy. One chapter describes life at Sling Camp.

Michael Morpurgo's children's novel, *Private Peaceful*, tells how Thomas Peaceful and his brother from a Devon village enlist and join 'a training camp on Salisbury Plain', where 'we slept in long lines of tents'.

Poetry

No other conflict – perhaps indeed no other event – has inspired so much poetry as the Great War and the names of many soldier-poets and celebrated lines from their works remain in the public consciousness. Not surprisingly, some trained in Wiltshire and in a few cases their work reflects their experiences in the county, as is the case with Francis Brett Young's 'Song of the Dark Ages', reproduced before Part One of this book.

Well before the cult of the war poet had evolved, Thomas Hardy's 'Channel Firing' refers to the sound of guns being heard:

> As far inland as Stourton Tower,
> And Camelot, and starlit Stonehenge.

This prophetic poem was written in April 1914. One may presume that Hardy chose the three locations because of their connections with British tradition: Stourton Tower being built in West Wiltshire in 1772 to mark King Alfred's victory over the Saxons in 879; Camelot being the legendary site of King Arthur's palace (popularly identified as being at Cadbury Castle in Somerset); and Stonehenge as the best-known ancient monument in the country.

At Marlborough College before the war, Charles Hamilton Sorley had enjoyed 'sweats' (cross-country runs) on the downs, often passing a signpost on Poulton Down, 2 miles north-east of Marlborough, which he mentioned in his poems, 'Lost', and 'I have not Brought my Odyssey', and in the first of 'Two Sonnets', originally titled 'Death – and the Downs', written four months before he was killed in action at Loos in 1915. Arguably his two best-known verses evocative of the war are 'All the Hills and Vales Along', perhaps written in August 1914, and 'When You See Millions of the Mouthless Dead', probably his last poem. Today in his memory a replacement signpost stands on Poulton Down (map reference 209717), with his initials carved in a slab of concrete beneath it.

On 2 September 1918, two days before being killed in a bombing raid on German trenches, Alec de Candole, another Old Marlburian, expressed his hopes that:

> When the long trek is over,
> And the last long trench filled in,
> I'll take a boat to Dover,
> Away from all the din;
> I'll take a trip to Mendip,
> I'll see the Wiltshire downs,
> And all my soul I'll then dip
> In peace no trouble drowns.
>
> Away from the noise of battle,
> Away from bombs and shells,
> I'll lie where browse the cattle,
> Or pluck the purple bells.
> I'll lie among the heather;
> And watch the distant plain,
> Through all the summer weather,
> Nor go to fight again.

De Candole had been a lieutenant in the 4th Wiltshire Regiment, before being wounded in October 1917. After convalescing, he was posted to a camp on Salisbury Plain (perhaps Lark Hill, where the 4th was then based) and wrote 'I Saw Them Laughing Once' and, just after Christmas 1917, 'Salisbury Cathedral'.

Not so fond memories were evoked by Lieutenant Edward Wyndham Tennant, born at Stockton House near Codford in 1897, his father becoming MP for Salisbury in 1906. A member of the 4th Grenadier Guards, he was killed at the Battle of the Somme in 1916. One of his poems reads:

> And here I am in Tidworth Camp
> By night I freeze, by day I tramp,
> We're crowded in a jolly squash,
> They never give us time to wash.
> The boys are dressed 'fore I'm awake,
> I need a dickens of a shake,
> But I don't mind the sweat and grind,
> For England, England's sake.

Other Old Marlburians who were war poets include Henry Field, author of twenty-six poems, who was killed in the 'Great Push' of 1916, and Arthur Jenkins of the Royal Flying Corps, a Balliol scholar who was killed in a flying accident on 31 December 1917, when with a home-defence squadron. Frank Morris wrote two sonnets about 'Belgium 1914' when still at Marlborough and died of wounds following his aircraft crashing on Vimy Ridge in April 1917.

Edward Thomas had several times visited Wiltshire before the war, including a period in 1907 when he was working on a biography of Richard Jefferies, one of the county's most celebrated poets. In 1915 he had written 'A Private', about a Wiltshire ploughman who dies in the war:

> This ploughman dead in battle slept out of doors
> Many a frosty night and merrily
> Answered staid drinkers, good bedmen, and all bores:
> 'At Mrs Greenland's Hawthorn Bush,' said he,
> 'I slept.' None knew which bush. Above the town,
> Beyond 'The Drover', a hundred spot the down
> In Wiltshire. And where now at last he sleeps
> More sound in France – that, too, he secret keeps.

'Bedmen' alludes to those who like to sleep in a comfortable bed. 'Mrs Greenland's Hawthorn Bush' is probably a joking reference to sleeping out of doors, 'Mrs Greenland' being the Downs personified. It is a hundred hawthorns that 'spot the down' above The Drover, which is a pub. As an officer cadet with the Royal Artillery, Thomas spent October 1916 at Trowbridge where he drafted two of his most celebrated poems, 'The Trumpet' and 'Lights Out'. He was killed in France on 9 April 1917.

The Wiltshire poet Alfred Williams was employed in Swindon railway works until 1914; he joined the Royal Field Artillery in 1916 and served in India, but only after the officer-in-charge at Devizes Barracks had advised him to return to the works, where he would be more use.

The troubled Ivor Gurney was a member of the 2/5th Gloucestershire Regiment at Park House Camp in 1916. Much of his life was spent in mental institutions and his first recorded breakdown was in 1913, but he appears to have been happy enough at Park House, enjoying the camaraderie.

Other poets of the Great War with Wiltshire connections include Captain Graeme West, who, as a private with the 16th Middlesex (Public Schools) Regiment, trained at Perham Down in late 1915; his best-known works include 'The Night Patrol' (March 1916), one of the first poems about the Front Line based on experience, and 'God! How I hate you, you young cheerful men!', a rebuttal of soldier-poets who were still writing about the war as a glorious game. He was killed in January 1917. Lieutenant Geoffrey Smith was at Codford with the 19th Lancashire Regiment in September 1915 and was inspired by a trip over Salisbury Plain to Westbury with Robert Gilson to write 'Song on the Downs' – his reflections on a Roman road across the Plain.

Smith was a friend of the author J. R. R. Tolkien and before the war, when a pupil at King Edward's School, Birmingham, had attended an Officers' Training Corps camp at Tidworth Pennings, where he scarred himself sliding down a tent pole to which a candle had been affixed with a knife. Gilson was with the 9th Suffolk Regiment at Sutton Veny Camp, and of his departure for France on 8 January 1916 wrote:

> I wish I could describe or draw for you the lovely sunrise we watched this morning from the train – like one of the Bellinis in the National Gallery, with Salisbury Plain standing up against the sky, banded by a lovely velvety black line – it is a long time since I have felt the sheer beauty of things so strongly.

John Streets, a Derbyshire coal miner whose pre-war poem, 'Truth, an Allegory', was well received by *Poetry Review*, enlisted in the 12th York and Lancaster Regiment (the 'Yorkshire Pals') in early September 1914. He wrote a number of war poems, including 'Sunset: Hurdcott Camp' and 'Hymn to Life: Hurdcott Camp', in which he contemplates death undaunted while exulting in life. Both poems are very spiritual. On 16 November, the 12th moved to Lark Hill, and in early December sailed for Egypt, from where it embarked for France on 10 March 1916. Streets, now a sergeant, continued to write poems in the trenches. He died in the great Somme assault on 1 July 1916. His collected works were published as *The Undying Splendour* in May 1917.

When the Australian Donald Clarkson was at Fovant Camp in 1918 he wrote a number of poems. One, 'The Sign Post at Salisbury', evokes the writer's thoughts as soldiers soon to leave for the French battlefields march past him. Its first two verses are:

> I stand aside at the old sign post
> and watch the troops go by
> rank after rank of an endless host
> whose footbeats reach to the sky.
>
> The light of youth on many a face
> the trace of a frown on some
> but they all march by with an even pace
> to the beat of the rolling drum.

Evelyn Southwell trained at Perham Down and Windmill Hill in 1915, when he wrote verse expressing his appreciation of the locality.

After the Great War, Owen Rutter, an officer in the 7th Wiltshire Regiment, wrote two long, acclaimed poems in the rhythm of *Hiawatha*. One, *The Song of Tiadatha*, describes an infantry subaltern's training and experiences of the Salonika campaign, the central character, Tiadatha ('Tired Arthur'), being 'a sort of composite of my brother officers'. The other, *The Travels of Tiadatha*, described the hero's adventures and wanderings after demobilization. After serving with the British North Borneo Civil Service, Rutter returned to Britain and was commissioned in the 7th Wiltshire on 12 June 1915, no less a person than Lieutenant-General Pitcairn Campbell, General Officer Commanding, Southern Command, certifying his good moral character. Rutter joined the 7th at Sutton Veny, where it had been since April, and it is probable that some of his verses relate to his experiences there. He embarked for Salonika on 21 September.

Poems of several of those mentioned above are celebrated as being amongst the finest of the war, but in contrast there were amateurish efforts best described as doggerel. Several companies produced postcards with verses describing conditions in camp, changing the name of the location with scant respect for scansion.

A poem about Sling Camp printed in the journal of the Channel Islands Great War Study Group has ten verses, starting with:

> Of all the places God hath made,
> Or human feet have ever strayed,
> I speak the truth, I'm not afraid –

HEYTESBURY CAMP.

THERE'S an isolated, desolated spot I'd like to mention,
    Where all you hear is "Stand at ease," "Slope Arms,"
    "Quick March," "Attention."
It's miles away from anywhere, by Gad, it is a rum'un,
A chap lived there for fifty years and never saw a woman.

There are lots of little huts, all dotted here and there,
For those who have to live inside, I've offered many a prayer.
Inside the huts there's RATS as big as any nanny goat,
Last night a soldier saw one trying on his overcoat.

It's sludge up to the eyebrows, you get it in your ears,
But into it you've got to go, without a sign of fear,
And when you've had a bath of sludge, you just set to and groom,
And get cleaned up for next Parade, or else, it's "Orderly Room."

Week in, week out, from morn till night, with full Pack and a Rifle,
Like Jack and Jill, you climb the hills, of course that's just a trifle,
"Slope Arms," "Fix Bayonets," then "Present," they fairly put you through it,
And as you stagger to your Hut, the Sergeant shouts "Jump to it."

With tunics, boots and putties off, you quickly get the habit,
You gallop up and down the hills just like a blooming rabbit,
"Heads backward bend," "Arms upward stretch," "Heels raise," then "Ranks change places,"
And later on they make you put your kneecaps where your face is.

Now when this War is over and we've captured Kaiser Billy,
To shoot him would be merciful and absolutely silly,
Just send him down to Heytesbury, there among the Rats and Clay,
And I'll bet he won't be long before he droops and fades away.

BUT WE'RE NOT DOWNHEARTED YET.

There were several 'stock verses' printed on wartime postcards, with the camp name included as appropriate.

> It's Sling
> It is no place to laze or lurk,
> Your duties there you cannot shirk,
>
> In fact I know you have to work
> At Sling.

It is thought that the author may be an R. Stanley, who may have been a Jerseyman.

One of the most famous poems of the Great War was written by the Canadian John McCrae, who arrived at West Down North Camp on

18 October 1914. He later wrote 'In Flanders Fields', which was to inspire the poppy as a symbol of remembrance. Its first verse is:

> In Flanders Fields the poppies blow
> Between the crosses, row on row,
> That mark our place; and in the sky
> The larks, still bravely singing, fly
> Scarce heard amid the guns below.

Another Canadian, Sergeant Frank S. Brown of Princess Patricia's Canadian Light Infantry, wrote a number of poems before dying on his first day in the trenches on 13 February 1915; edited by Holbrook Jackson, they were published as *Contingent Ditties – Other Soldier Songs of the Great War*. Jackson noted in the preface that 'at the time of our correspondence [presumably in late 1914] his regiment had proceeded to France, but the soldier-poet, much to his disgust, had been left behind, temporarily unfit through an influenza attack. His first letters to me were written from Lark Hill Hospital, Salisbury Plain.' (Actually there was no hospital at Lark Hill then and Brown was probably at Bulford Manor Hospital.) On discharge, Brown reported to the short-lived Canadian depot at Tidworth, before leaving for France.

The author and poet Siegfried Sassoon (another Old Marlburian) bought Heytesbury House in 1933. As well as his distinguished works he is remembered for his protest against the conduct of the war, which included sending a letter, 'Finished with the War: A Soldier's Declaration', to his commanding officer and throwing his Military Cross ribbon into the River Mersey.

11
A Slow Return to Peace

Unrest and Demobilization

As soon as news of the Armistice of 11 November 1918 reached the military bases in Wiltshire, the reaction after the initial delight and relief was 'How soon do I get demobbed?' In a great number of cases it would not be nearly fast enough and very soon there was unrest and trouble in many camps.

Days after the Armistice, morale among a Meteorological Section of Royal Engineers at the School of Navigation at Stonehenge Airfield plummeted when a staff major from the War Office asked for volunteers to join the North Russian Expeditionary Force; only one man came forward. Among those most horrified by the prospect was Lance-Corporal Andrew Rothstein, who went on what he hoped would be an extended Christmas leave only to receive a telegram ordering his immediate return. He was one of seven men who were to form a compulsory draft for Russia.

Then came news of unrest among troops at Dover and Folkestone and on 4 January Australians at Lark Hill ran amok; next morning there was a wreath of smoke above the camp, the military cinema being among the buildings burnt down. Soon every airfield on Salisbury Plain had elected, or was preparing to elect, a protest committee. A corporal in Rothstein's unit declared 'we've had enough of this bloody war and this bloody army'. Rothstein himself took the view that the war in North Russia was undeclared and illegal, and being posted there would be contrary to the undertaking of his enlistment, which was to serve for 'three years or the duration'. But by the time he saw his commanding officer on the 6th the Government had heeded the swell of protest and he was told that demobilization would start four days later. (In fact, the war was not yet over, as only an Armistice had been agreed and the Treaty of Versailles to end it formally was not signed until June 1919.)

At Lark Hill in February 1919 twenty or thirty members of the Maori Pioneer Battalion, some with pistols, broke into the canteen and stole a 36gal cask of ale and rolled it down Amesbury Road. At Sling on 14 and 15 March New Zealanders, angered at the delays in going home, raided messes, stealing cigarettes, beer and food and destroying furniture and furnishings. Four sergeants and various privates were court-martialled for 'endeavouring to persuade persons to mutiny' and 'in joining in a mutiny'. Three of the four sergeants were reduced to private and sentenced to up to six months' imprisonment and hard labour; privates received up to 100 days.

Much press coverage was given to riots in towns, though judging from local newspapers the only serious trouble away from the camps in Wiltshire was in Swindon, where peace celebrations held in late July 1919 were marred by ex-servicemen upset about their lack of special recognition in the festivities. Some may also have been disgruntled by modest gratuities, a housing shortage and poor job prospects. A pole bearing a 'Flag of Peace' was burnt down – though the flag itself was saved – after rumours had spread that it was to be the town's only memorial to the dead and that it had cost £200. The *Swindon Advertiser* and local officials tried to avoid blaming former soldiers and sailors, noting that some had formed pickets to maintain order and peace, and claiming that 'hooligans' of unspecified origin were to blame. However, troops from Chisledon Camp were seen to have been involved.

Censorship regulations (which remained in force for several months after the Armistice) prevented local newspapers reporting the disturbances within camps. But at many of these there was continuing unrest among Dominion troops awaiting repatriation, which was due particularly to frustration over delays in getting home caused by a shortage of shipping. Some who did return home early in 1919 complained of unhealthy conditions and bad food on their voyage.

With an uncertain civilian life awaiting them, many soldiers still in England resented the continuing military discipline, though there were attempts, such as those by the New Zealand and Australian authorities, to provide training in civilian occupations. New Zealanders at Sling were put to work carving the shape of a giant kiwi in the hillside to keep them occupied. Some Australians deserted and lived rough in the woods near Fovant, and in July there was a mutiny, leading to courts martial, among the 3rd West Yorkshire Regiment at Durrington Camp.

Fovant and Chisledon camps were two of seventeen demobilization camps in Britain that had a target of processing 40,000 discharges a day, a figure some thought would be difficult to attain. Henry Prittie devoted a chapter of his autobiographical *Khaki and Rifle Green* to his command of the demobilization unit at Chisledon Camp: 'Arriving there I found a scene of indescribable confusion. About a hundred men had turned up [as part of his unit] and a pretty rotten lot they were, all of considerable age and none of them had been overseas.' For guidance, he was supplied with a demobilization book from Japan (hopefully translated into English), as that was the only country to have disbanded a more or less civilian army after its war with Russia in 1904–05.

A Slow Return to Peace

This card symbolizes Australians' frustration at delays in returning home in 1919. Having crossed the Channel from France, they find that their journey back to Australia is much delayed, hence the 'gates' that are confining them to Fovant.

Prittie's teams worked from 6am to midnight and, he says, once discharged 10,000 men in a day with time to spare, compared with a War Office estimated target of 3,000. High-placed visitors never quite believed that men reaching the camp by 10am would leave it by 4pm and insisted on going through the process for themselves. Steel helmets were popular souvenirs, though when one visitor thought the men would like to retain their tattered greatcoats as mementos there were no takers. The reason emerged when civilians complained that men arriving by train had left 'a lot of little things behind them', resulting in a delousing station being set up. Then church leaders thought it a shame that the returning heroes had to go home in rags, so a reclothing section was instituted to issue them with clean khaki (which, Prittie thought, would soon be sold for what it would get).

Then Prittie's own 'home servicemen' (those who had never served overseas), composed of 'fathers and grandfathers', mutinied, wanting priority in the return to civilian life. He had a word with the major commanding the camp's machine-gun school:

> Then I went out onto the porch of the orderly-room. About a thousand men were howling: 'We're going to be demobbed first.' A note on the bugle and then a dead silence till I took up the tale. 'There are Lewis guns in position commanding every street. My signal on the telephone and they open fire. Ten seconds to get to your huts.' The allowance of time was over-generous. In 5 seconds every home Serviceman was under his cot. I walked around with my staff and made the necessary arrests. Those who had yelled the loudest in front of the orderly-room.

As a precaution Prittie arranged for a squadron of Reserve cavalry to come up from Tidworth, followed next morning by a train-load of armed riflemen.

However, he sided with the men when they complained after their rations were drastically cut because they were deemed to be performing 'sedentary duties'. He told the War Office that 8,000 soldiers were expected the next day and might take their revenge on Swindon if his unit was not functioning. Full rations were quickly reinstated.

One day Prittie heard from Horatio Bottomley, editor of the pro-soldier magazine *John Bull*, that a Labour Member of Parliament had received a complaint about a constituent who had been demobbed at Chisledon. It was claimed that the man had been kept waiting in snow and sleet for eight hours, had been jammed for the night into a leaking tent with twenty others, had not been fed, had stood about next day until he was turned out of camp at nightfall without food or railway ticket, and on finally reaching home had died from pneumonia. Prittie replied that there had been no snow and, as the men were in huts, no tent, the private had spent two hours in camp, half of the time being spent consuming a large hot meal, before departing with ticket and gratuity, but he had taken three and a half days to find his way the 60 miles to home. Bottomley, who had known Prittie before the war, accepted this version.

Guy Mace of the Royal Engineers was less impressed when he arrived at Chisledon at 3.15pm on 24 February 1919, finding 'utter chaos ... no one knew anything or what we were supposed to do'. After an hour or so Mace and his comrades marched to a large shed, where they handed in certain items of kit and their rifles and had 'a meal of sorts'. He finished the demobilization process at 7pm, but was not allowed to leave camp until he caught the 10pm train for Swindon that departed an hour late.

A Slow Return to Peace

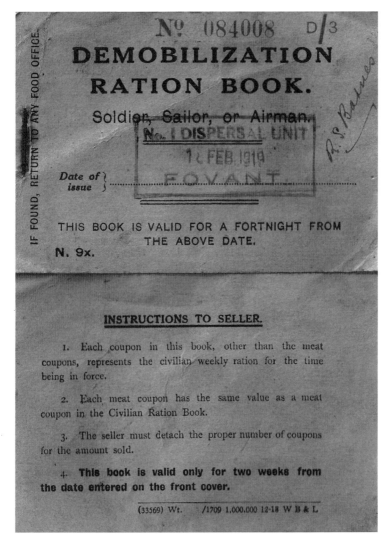

Ration books were issued to soldiers at demobilization centres such as Fovant. This one was given to Sergeant R. S. Barnes of the East Kent Regiment and still contains coupons for tea, sugar, meat, cheese, jam, lard, butter and margarine.

One lieutenant managed to be issued with a warrant to travel from Swindon to Edinburgh via Penzance! The War Office had recently refused Prittie's request for a Bradshaw railway timetable, a map of the British Isles and regional railway guides, and this was pointed out to higher authority when it

A Slow Return to Peace

queried the roundabout journey. The requisite publications arrived shortly afterwards.

In July 1919, 357 absentees from New Zealand forces (who were based mainly in Wiltshire) were recorded and newspapers carried warnings that all NCOs and other ranks should report to Sling Camp on or before 31 July or be struck off the strength of the Expeditionary Force, thus forfeiting passage home, war gratuity and any other privileges granted to NZEF soldiers. A similar warning was given to Australians, but with a deadline of 15 August. The process of demobilization continued well into 1920, with there still being more than 5,000 Australians in Britain in February.

By the end of April that year all British warrant officers, NCOs and other ranks who had enlisted for the war's duration, or who had been called up under the Military Service Acts, were back in civilian life (except for those who had volunteered to stay in the Army). The dispersal unit at Fovant Camp had been disbanded on 15 January, with men who had served overseas and were still to be demobilized in south-east England then going to Purfleet in Essex. A rest camp was established at Fovant for drafts disembarking at Southampton and Devonport and due to be demobilized elsewhere. This rest camp remained until June, when the drafts then went to Shorncliffe.

Aftermath of War

For many thousands of men, Fovant and Chisledon camps provided the last taste of the Army, though others were destined for further wartime service twenty years later. Indeed, a few Reservists were recalled in 1921, when miners and railwaymen went on strike. Government action included mobilizing Royal Navy stokers to look after the coalpits and pumps, and then soldiers to look after them after they had faced intimidation from the strikers. This led to the unusual presence of sailors at Tidworth Barracks, where they spent several weeks in case they needed to be sent to trouble spots.

For soldiers returning home to Wiltshire there was a warm welcome, with many towns and villages hosting receptions and dinners for them. But demobbed soldiers at Dinton were disappointed that they had to organize and pay for their own dinner, when no offer was forthcoming from their community. Over the next few years there were sombre reunions, not only on Armistice Day but at ceremonies to unveil memorials to those who had not returned. (As early as 6 October 1917 the Bishop of Salisbury had unveiled a memorial to the men of Australia's First Training Battalion in Durrington Cemetery.)

By 1920 Ludgershall had three war memorials, each giving a different year for the end of the war: 1918 (when the Armistice was declared); 1919 (when the Treaty of Versailles peace document was signed); and 1920 (when the treaty came into effect). Salisbury's memorial was unveiled in 1921 by Lieutenant Tom Adlam, born in the city in 1893 and who, as a member of the 7th Bedfordshire Regiment, had won the Victoria Cross. Soon every village had its roll of honour giving the names of the dead, often in such numbers as to make one ponder the effect a war fought overseas had had on the most peaceful of communities. Sometimes the same surname appears twice or more, showing the scale of sacrifice made by one family.

At Marlborough College, 320 of whose old boys had died in the war, there was not much support for the original idea of a memorial cloister, the mass of opinion favouring a hall close to the chapel, costing £35,000. The result was a semi-circular building with wooden seating for 1,516 and with the names of the fallen inscribed on the walls. At Dauntsey Agricultural School a mural tablet of teak was unveiled in 1920 to commemorate the thirty-seven old boys who had died in the war.

In a letter published in *The Times* of 28 September 1918, H. D. Rawnsley suggested a novel way of remembering the dead:

> Salisbury Plain as a military training ground has done much towards winning the war. What could be more fitting than that here, in the midst of Salisbury Plain, there should be at this old meeting-place of prehistoric tribesmen and warriors, an assemblage on Midsummer Day of each year, or at stated intervals: and that a solemn service should be held in memory not only of Wiltshiremen but of all the men of the British Empire who have died for right against might – for justice, freedom and peace? ... Nothing would be needed but a huge stone Celtic cross, in the neighbourhood of the circle, with a simple dedication to the imperishable memory of the gallant dead.

The idea was not well received; indeed, a Canadian battalion had been allowed to hold a church service at Stonehenge in January 1915 only on condition that it was not to be a precedent. Now there were deep concerns about how to preserve the monument and crowds of people near it would not have been welcome. For some years, several of the stones had had to be supported by wooden props that stayed in place until a restoration programme in 1919–20. But anxiety continued about the nearby airfield dominating the skyline and how to protect the site from both the public and the latter-day Druids who wanted to bury their dead there.

Over the next few years the Imperial War Graves Commission was busy all over the world ensuring that the bodies of service men and women were permanently interred with dignity, the original wooden crosses being replaced by the white stones that have become iconic. At one time, 1,971 military graves of the Great War period were located in 196 Wiltshire burial grounds, with 1,114 concentrated in six locations: Bulford, Codford St Mary, Durrington, Fovant, Tidworth and Sutton Veny. Included were those of some Americans,

The unveiling and dedication of the Cross of Sacrifice at Sutton Veny on 22 July 1923. The graves of soldiers with the original wooden crosses can be seen.

whose bodies were later removed, and those of more than 100 Germans, mostly at Durrington, who would be reburied at Cannock Chase. The burials at Durrington are in the cemetery next to the A345 (map reference 153453), where there is a memorial to the Australian Imperial Force and a plaque commemorating the First Canadian Contingent. Each year on or close to ANZAC Day (25 April) ceremonies are held in several Wiltshire churchyards to remember those Australians and New Zealanders who died so far from home.

Many survivors of the war needed special help to return to civilian life because of injury or ill-health. In January 1919 the Village Centres Council confirmed that it would open 'a medical treatment centre and training depot' at Enham in Hampshire, 2 miles north of Andover, providing education in horticulture, agriculture and carpentry. A thousand acres were purchased for £30,000, with places for 150 men suffering from shell shock, neurasthenia and depression, and physical injuries such as the loss of limbs. Those recovering from malaria and infectious diseases would also be admitted. The total cost of purchasing and equipping the centre was put at £100,000, and the aim was to treat 1,000 men at any one time. It opened in June.*

---

*After the Second World War 'Alamein' was added to the settlement's name, following a donation from the Egyptian Government and in recognition of the famous battle. Since then, Enham Alamein has continued to rehabilitate and train disabled people.

Disposal of Surplus Material

After the war the Army found itself with half a million surplus huts and vast amounts of unneeded materials and stores, the disposal of which ensured much profitable business for auctioneers and dealers. Many camps that had stood for only four or five years quickly became ghost towns and several airfields that had been built only in 1917 and 1918 became redundant. In *Goodbye to All That*, the poet and author Robert Graves recalled how he and his wife, when cycling down to the West Country in 1920, 'rode across Salisbury Plain, passing Stonehenge and several deserted army camps, which had an even more ghostly look. They could provide accommodation for a million [*sic*] men: the number of men killed in the British and Overseas Forces during the war.'

Nearly every week for two years and less frequently after that, the advertisement pages of Wiltshire papers carried notices of sales, with the name of Salisbury auctioneers Woolley & Wallis being prominent. Stores at the wartime airfields were among the first to go, early in 1919, but soon massive sales were being held at all the camps. After the Auckland Detachment of the New Zealand Expeditionary Force left Lark Hill, its grand pianos, billiards tables and sports goods were on offer. In 1920 buildings were being sold off in large lots, for example 250 huts from Park House and an initial 300 huts from Chisledon in April, and then a further 189 in July. In April stores for sale at Rollestone Camp included 717 chairs, 11,075 blankets, 3,780 palliasse covers, 315 galvanized urine tubs and 1,861 wire coathangers. At Tidworth 450 Canadian wagons were for disposal. In June 28,731 serviceable blankets were on offer at Corton Camp. The advertisements also reveal the effect the diminishing military presence had on businesses, such as in May 1920 when the contents of the Victory Café at Shipton Bellinger were for sale, its soldier patrons having left nearby Park House Camp.

'The auction sale of Australian motor cars at Salisbury Plain was highly successful,' reported the *Brisbane Courier* of 20 September 1919. 'The same may be said of the typewriters, which realized above the cost price. Much Australian clothing at recent sales by auction was purchased for Norway, Italy, Jugo-Slavia and Persia.'

Sales went on for several years. On 30 January 1922 Harding & Sons – auctioneers based in Warminster and Frome – were offering eighty-four huts and other items on site at Codford, Corton, Heytesbury and Upton Lovel. The huts varied in size: one was merely 6ft by 6ft 6in, another was 190ft by 28ft and comprised five rooms.

Many towns and villages acquired relics of war, such as German tanks and guns. At Marlborough a gun on display near a busy road proved attractive to children, who were thought to be at risk from traffic. The War Office presented

Marlborough College with a 77mm German gun in recognition of the services of its Officers' Training Corps, 1,419 members of whom had joined the armed services between November 1908 and July 1918. The college also received another gun from the War Office, a 5.9in howitzer from the Lord Lieutenant of Wiltshire and an old English carronade captured from a Turkish fort by members of HMS *Marlborough*.

The War Savings Committee presented some 260 'war-battered' tanks to towns and cities, including Salisbury and Trowbridge, that had raised the most money by their residents' purchases of National War Bonds and War Savings Certificates. British Mark IV tank 222, which had served in several 1917 battles, was delivered to Trowbridge on 19 December 1919 and set up on what was later the war memorial site; when this was built in 1921, the tank was moved to nearer Castle Street and, as with other presentation artillery and tanks, in 1940 was cut up as part of the war effort. Tank 211 was presented to Salisbury.

There were suggestions that some huts, such as those in the Wylye Valley, could remain to form civilian villages, but in the event most were put up for sale, dismantled and removed to other sites. It was estimated that a hut priced at £100 could be converted into a home at a cost variously put at £190 or £300, including dismantling, transport, re-erection and partitioning into sitting rooms and bedrooms, larder, scullery, bathroom, coal house and 'other conveniences'. But many local councils thought the huts too primitive to allow their use as homes, despite an acute housing shortage. However, Liverpool Council did buy 500 from Lark Hill Camp for conversion into family houses, each comprising a living room, three bedrooms and scullery. Others were sold off to local farmers and owners of small businesses. Swindon bought one for £65 and erected it in the girls' playground at the Secondary and Technical Institution; other schools to do likewise were those at Trowbridge and Bradford-on-Avon. The Board of Agriculture and Fisheries also expressed interest in acquiring ninety-four huts to house prisoners of war employed on drainage schemes, although eventually it concluded that none was suitable after the costs of transport to new sites were taken into account. Many huts also became village halls. The celebrated local writer Alfred Williams acquired some 'very good timber from Chiseldon Camp' and built a house at nearby South Marston. By February 1924 Britain had realized £680,000,000 from the sale of surplus war goods worldwide, but still there was a need to retain a massive War Department stores depot at Swindon.

The YMCA and other welfare bodies also had to dispose of their huts and in several cases seemed to have asked high prices. This was the feeling of the Ludgershall Comrades of the Great War, who wanted one as a clubroom. They protested to the YMCA National Secretary, Sir Arthur Yapp, which led to a visit by officials who agreed more reasonable terms. The YMCA hut in

Bridge Street, Andover, was moved to Enham, to the disappointment of local lads who had been using it as an alternative to walking the streets. Again, there seems to have been some haggling over the price. The hamlet of Bemerton also acquired a YMCA hut.

One hospital hut was moved from Salisbury Plain to the hamlet of Woodmancott, between Basingstoke and Winchester, to form the basis of a residence. Much added to, it came on the market in the winter of 1999–2000 for £324,000, a price that reflected the large area of land on which it had been rebuilt.

Where it had commandeered farmland on which to build camps, the War Office offered £30 an acre compensation before returning it to its owner, but this was insufficient to encourage some farmers to remove concrete foundations and they contented themselves with cultivating the areas in-between. In the short term, this was perhaps the most sensible financial solution, given the post-war state of farming, but it left some unsightly scars to be cleared away one day. This entailed breaking up the concrete with sledge hammers and piling it along hedges, with much of it eventually being taken away as hardcore for new developments. Here and there, some remains can still be seen, such as brick and concrete cubes to do with drainage abandoned on the edges of 'Hospital Field' at Fovant.

Certain structures that most people wanted removed as soon as possible remained for a decade. They were on the airfield built in 1917 on leased land 600yd from Stonehenge. Even allowing for wartime exigencies, it is surprising that the authorities chose this sensitive site, the more so as there had been great public concern about the ancient monument when the Amesbury estates of Sir Cosmo Antrobus were auctioned in 1915. (Stonehenge was bought for £6,600 by Sir Cecil Chubb, who later presented it to the nation.) On 27 August 1918, describing the view from a point to the east of the road fork, a special correspondent wrote in *The Times*:

> On one side a great camp, very high, dominating buildings, spreading, spilling themselves down the slope. On the other a secondary camp, cresting the skyline. And between them, on the lower ground, a little huddle of dark, insignificant objects – can that be Stonehenge? They used to look like monstrous blocks which the children of the Giants had begun to play with, and abandoned in their play. Now ... they are dwarfed, obsolete, unmeaning.

On 20 December 1919 the Society of Antiquaries of London sent a strongly worded letter to the Government about 'the complete disfigurement of the surrounding of the most famous of British monuments'. An official, from the date of his letter apparently working on Christmas Day, replied that the:

> Air Council are in full sympathy with your Society's desire that existing disfigurements in the neighbourhood of Stonehenge shall be removed at the earliest

A Slow Return to Peace

possible moment & would be very glad to vacate the aerodrome at once and allow the buildings on it to be taken down, if a possible aternative [*sic*] site existed.

However, the letter continued, the site was of value because of its proximity to Lark Hill and other artillery camps and the war had proved the necessity for cooperation between aircraft and artillery. No other aerodrome was sufficiently nearby to facilitate the appropriate training on the Plain. The following June advertisements appeared for the sale of 'Stonehenge Night Camp', which unusually stated that some of the buildings, including an aeroplane shed, were 'not for sale for removal', and suggested they were suitable for use for laundry or storage purposes. The stipulation that they could not be removed is not only curious but seems insensitive, especially as many other military sites in Wiltshire were being cleared and returned to their original state. (Indeed, the county fared better in this respect than Regent's Park and Victoria Embankment in London, where wartime huts were not removed until 1924.)

But a year later alternative arrangements for training in army–aircraft liaison had been made, with the locating of the School of Army Co-operation at Old Sarum Airfield, and the Stonehenge buildings passed into the hands of the Disposal Board. In February 1922 these were sold by auction on the understanding they would be removed, but the purchaser sold them on to the owner of the land. The Government received legal advice that the latter could not be compelled to remove them. They were used as a pedigree stock farm

Taken in 1927, this photograph shows how the bomber hangars of RFC Stonehenge dominated the ancient site – and how the new building in the foreground also marred the landscape.

and living accommodation before coming on the market again in 1927, when there were fears that they might be replaced by a 'bungalow town'. (Intrusive buildings, including a café, had recently been built to the east of Stonehenge.) At the same time that a local preservation fund was set up, a national campaign was launched to raise £35,000 to purchase the land for a mile around Stonehenge, with George V heading the list of subscribers. In the event, this time the buildings were sold 'for removal', and their demolition started in October. A year later Charles Palmer of Shrewton was breaking up concrete foundations when he discovered what some experts thought might be an ancient Egyptian scarab. It proved to be a modern replica sold to tourists in Egyptian bazaars, perhaps a souvenir left behind by an ANZAC soldier who had served in Egypt before coming to Wiltshire.

Back to 'Real Soldiering'

It took several years for Army life in Wiltshire to return to 'real soldiering', to use the term popularly attributed to Regulars who regarded war as an interruption to rigid discipline, faultless ceremony, immaculate barracks and occasional manoeuvres. Before this was to happen, all the barracks and camps became increasingly depressing as the numbers of soldiers dwindled, stores were auctioned and the buildings allowed to deteriorate before, in most cases, being sold and removed. The *Andover Advertiser* describes a particularly gloomy scene at Perham Down Camp, periodically reporting the dereliction, the effect the absence of soldiers had on Ludgershall traders, and the uncertainty about the camp's future. (But it is worth noting that the villagers were not sorry to see the Royal Inniskilling Rifles leave Tidworth in October 1919 after they had caused trouble in Ludgershall, one resident being wounded by a knife.) Ironically, Perham Down was one of the few wartime hutted camps to have a peacetime role.

In May 1919 the Crown Prince of Denmark inspected artillery units at Lark Hill and Chapperton Down, being quickly followed by a Spanish military mission. No doubt these visits reflected the international realization that artillery had a crucial role in modern warfare and Lark Hill (or 'Larkhill', as it was now more usually spelt) went on to establish a worldwide reputation for its expertise, training members of many foreign armies.

Officers' Training Corps camps returned to Tidworth Park in July 1919; a year later there were up to 6,000 cadets at Tidworth Park and Tidworth Pennings, but no large Territorial camps were held anywhere in the country. The Territorial Force had been disbanded in 1919 but was reformed the next year as the Territorial Army, with many battalions being under-strength. As before the war, employers who were reluctanct to release workers for Territorial summer camps were blamed and some artillery units were even using German guns because of shortages. There is a suspicion of kindness about the venues for the small Territorial camps that were held: Scarborough, Paignton and Shoreham, for example; all 'soft' localities compared with the Plain. *The Times* remarked in 1921 that summer camp had provided many recruits from the poorer parts of London with their first glimpse of the sea. The following year it noted that the Beaulieu area of the New Forest was a popular Territorial camping-ground and that Shorncliffe offered 'all the attractions of the seaside'. (And in 1925 it was to lament the lack of a cinema during the Territorials' summer camp at Beaulieu, suggesting that if the weather was bad then the bored soldiers would give a poor account of their experiences on returning home.) How could Salisbury Plain compete, with its natural bleakness, derelict hutments and years of wartime detritus?

In 1920 Tidworth Barracks was still only half full, and even when 750 men of the Loyal North Lancashire Regiment arrived in late August many immediately went on leave prior to going overseas. However, the *Andover Advertiser* accurately predicted in September the arrival of tanks on Salisbury Plain and soon a new era of military training began.

It was not until 1924 that Salisbury Plain saw manoeuvres anything like those of before the war, with Fargo, Bustard and Bulford Field camping-grounds being used by Territorials and the Regulars holding modest exercises on the Plain, notwithstanding a late harvest that limited the amount of land available to them. Cavalry were still evident, but so now were tanks, armoured cars, the Royal Air Force and Admiral Sir Sydney Freemantle from Portsmouth, whose presence confirmed full inter-service cooperation. However, the Army was aggrieved at its lack of funding compared with the Senior Service and artificiality and improvisation remained features of its training activities.

Army chiefs were looking forward to 1925 and manoeuvres on a pre-war scale, and so it proved when a large 'expeditionary force' assembled in Hampshire and Wiltshire on 17 August, the start of a month of full activity in the area. The directing staff established their headquarters near Wylye and there was a tented camp near Sutton Veny, not on one of the wartime sites but to the west, at Long Iver. 'Northesk' force headquarters were at Codford and the opposing 'Southesk' troops were prominent around Fovant. Steel helmets, artillery manifestly more mobile than before the war and an air force liaising with ground troops all spoke of a new era. The advent of tracked and other powered vehicles added to noise, dust and road damage.

Whereas in the early twentieth century many had seen the war as inevitable, there were now those who believed that it had been the one to end all wars. But in Wiltshire's barracks and on its training grounds the preparation for armed conflict continued.

PART TWO: THE CAMPS

BOYTON AND CORTON

Sites for hutted camps were chosen at Boyton and Corton near Warminster soon after the outbreak of the Great War. They had had a brief foretaste of military occupation four years before when manoeuvres were held in the area (as they had been in 1898), with a hundred soldiers of the Royal Scots Greys being billeted in the outbuildings of Sundial Farm, Corton. Field Marshal the Duke of Connaught and six other officers stayed in the farmhouse and fifteen members of staff camped on the lawn.

Recruits from Kitchener's Army who arrived in the locality in September 1914 had no uniforms or change of clothing for several weeks and lived under canvas. Among them were 480 men of the 8th Oxfordshire and Buckinghamshire Light Infantry, who moved into tents in Boyton Park on 14 October. Numerous of these were soon discharged as unfit, despite having been accepted by the regimental medical officer when the battalion was formed at Oxford. The 8th remained at Boyton for a month, during which little training was possible because of the weather. On 16 November it thankfully moved to billets in Oxford, at the same time as many other units camped nearby were similarly accommodated, mostly in counties adjoining Wiltshire.

The building of huts, mainly to house artillery and veterinary units attached to divisions quartered in nearby camps, started early in 1915. Not many references to activities there survive, and some of those that do suggest the two sites had separate identities, others that both were known as Boyton Camp, which would have been the more logical, as Corton (or Cortington) was a hamlet within the parish. A plan of 'Boyton Camp' made in May 1915 embraces Camp 1 in Boyton Park and four more groups of huts (three of which are numbered 2, 3 and 4) 500yd away, close to Corton. The Boyton Camp postmark was applied to mail from all of these. On the other hand, there are many references and photographs specifically naming Corton Camp, the larger of the two sites. A 1917 list of 'War Department occupancies' refers to 'Boyton and Corton hutment camp'. It is likely that Boyton was the preferred name to avoid confusion with Corton in Suffolk.

The diary of the 38th Mobile Veterinary Section of the 26th Division records that it left Corton at 4.15am on 21 September 1915 to entrain at Warminster, arriving at Southampton at 9.30am for embarkation to France. Two years later an Army veterinary hospital was noted at Corton.

With more activity in much larger camps nearby, Boyton and Corton seldom featured in the local press, the most prominent coverage being when a local dealer and an Army lieutenant appeared in court in 1916 charged with the unlawful disposal of Government property from Boyton. A quartermaster-sergeant was also charged with the unlawful disposal of horseshoes.

Despite the misspelling, this cards shows Corton Camp in 1915. 'It looks very nice on a Post Card but you should see it after a little rain,' writes the sender. 'We mobilize the 10th of Sept and expect to leave the 15th Sept most of us go for some Gun firing for several days.'

Boyton was one of the first wartime camps to be dismantled, with local newspaper advertisements in February 1919 offering all the materials there for sale. 'Huts could be adapted for dwellings, farm buildings, warehouses, public halls, stables etc,' prospective purchasers were advised. 'As it is probable that this is the only camp in the neighbourhood which will be dismantled within the next 18–24 months, any one wanting these materials should apply at once.' A map printed in 1920 shows huts still standing only at Corton (and none at Boyton) and one of 1923 reveals hardly any trace of these. In 1922 several large buildings had been auctioned on site, including officers' quarters, two dining-rooms and a double horse-shelter and forage store measuring 203ft by 28ft.

BULFORD

Having satisfactorily used land near Bulford as a camping-ground during its 1872 manoeuvres, the War Office completed its first purchases of acreage there in 1898, but had access earlier than this, with the 4th Cavalry Brigade camping on Bulford Down in July of the previous year.

In 1900 troops, including the 12th Hussars, were based there under canvas until early November, when, following several deaths from pneumonia, they were dispersed to elsewhere in England and Ireland. By then, John McManus had begun constructing brick barracks at Bulford. Progress was rapid, despite an infestation of rats in 1901, and in July that year work was aided by 200 men of the Monmouthshire Engineer Militia.

The *Daily Express* in 1906 described the barracks as 'a huge fungus, looking for all the world like a Western mining camp grown into town-like proportions over night'. The *Andover Advertiser* of 18 May that year noted that: 'the isolated position of Bulford has given it a terribly bad name amongst officers and men, who sadly bemoan their fate at being stationed there, especially in winter time'. Two years later, petty offences and drunkenness were reported among the Royal Scots Greys, said to be fed up with the isolation of Salisbury Plain.

A particular complaint among married officers was the initial lack of suitable family housing and even senior officers' wives had to live in tiny tin shacks. But the larger buildings, including warrant officers' quarters, were built of brick, and K. R. Stead of the Denbighshire Hussars Imperial Yeomanry was being unfair in 1909 when he wrote: 'This is a tin city, all buildings of corrugated iron'.

From 1903 to 1906 an instruction school for mounted infantry was based at Bulford, its existence reflecting a short-lived enthusiasm for such troops after the Boer War. But early in 1906 came the decision to rationalize such schools nationwide and that at Bulford was closed, leading to speculation about the future of the half-empty barracks.

As early as 1910 buildings were being demolished and replaced with new ones. The removal of a stable block and the building of married quarters nearby were signs of the time – less dependency on the horse, more awareness of the Regular soldier as a family man.

With the outbreak of the Great War, Bulford became a major Army Service Corps depot and further accommodation was built, including hutments near Sling Plantation, which became identified as a camp in their own right. The Bulford Mobilization and Embarkation Area opened in September 1914 to issue and repair vehicles and despatch them and men overseas. The Workshop and Stores Department received up to a hundred vehicles a day from civilian factories or from overseas, at any one time holding 700 under cover.

Visiting Salisbury Plain in late 1914, Rudyard Kipling noted the use of the term 'Bulfords' to refer to vehicles supplied by the ASC when a corporal whose unit had received such motors 'outlined the more virulent diseases that develop in Government rolling-stock (I heard quite a lot about Bulford)'.

On 14 August 1915 the *Evening Post* of Wellington, New Zealand, printed a letter from Arthur Chorlton who had joined the British ASC and was at Bulford. Chorlton reported that half the motor vehicles there were American,

A variety of wagons at the Army Service Corps workshops at Bulford.

either Packard or Peerless, and the others were Daimler, Dennis, Albion, Leyland, Commer, Halford and half a dozen other makes:

> Drivers are now tested on our parade ground in a complicated figure of eight course, laid out by faggots and logs on each side. You start up, run up the hill, changing gears to suit getting into 'top' on top, and then coming down for a trick test on the down grade. You go round the next half of the figure of eight, and then backwards through a circuitous track bounded by more logs. This reverse is the chief difficulty, as I found, but I managed to pass my test satisfactorily. Some of the men were quite novices, and the tests were as good as a cinema freak film. I christened the place the Bulford Motor Circus.

As at most Wiltshire camps, Australian units were based at Bulford from mid-1916, with Number 1 Australian Dermatological Hospital taking over existing facilities and treating cases of venereal disease, at one time being able to accommodate 1,520 patients. Some of these were under military detention for crimes and needed a guard (typically of one officer and seventy-eight men), who proved unable to prevent escapes from an insecure building. Eventually criminal patients were treated at Lewes Prison.

Food was very basic, as at all camps (except at Christmas and just before a unit left for overseas, when special efforts were made), and became more so as the war progressed. Corporal William Beer wrote to his mother in September 1918:

> The latest item in the messroom is the return of the Dog Biscuit to substitute [for] bread. Last winter we had them but you could masticate them. These

Bulford

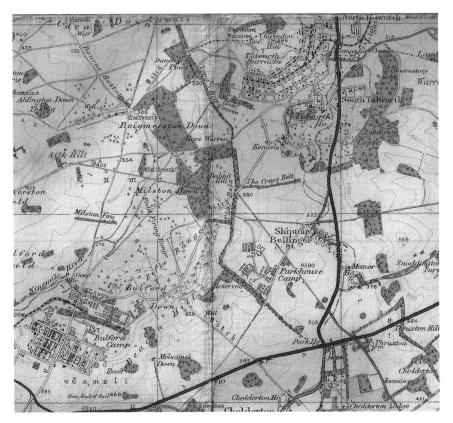

This extract from a 1in Ordnance Survey map of 1920 shows Bulford and Tidworth Barracks and, in-between them, Park House Camp. (Not named is Sling Camp, the northern part of 'Bulford Camp'.)

biscuits however are as hard as stone with hardly any flavour & if you do not want to stay for an hour or more at the meal one has to let little pieces of biscuit slip down which by no means aids digestion. The 'WAACs' [Women's Army Auxiliary Corps] are having them as well & they have fairly got the 'wind-up'. However they are privileged to have one piece of bread as well, we have 2 biscuits and no bread.*

---

*The introduction of such biscuits at Chisledon Camp led to the stamping of hundreds of hobnailed boots in the dining-hall and shouts of 'we want bread, we want bread'. Next day, there were no biscuits but a slice of bread for each man.

On 26 September Beer wrote home again expressing the indignation of soldiers at Bulford about a railway strike. Feelings were high, 'especially as hard biscuits have been issued lately, while they [the railwaymen] are asking for more money'. He mentions an appeal for soldiers who had worked on the railways in civilian life to come forward, but the strike ended before their experience could be put to use.

In a letter of 5 November Beer noted a new addition to the office where he worked, a black man: 'a West Indian ... soldier. He is only here temporarily & and is supposed to be a clever man in the West Indies having the degree of Batchelor [sic] of Science. He is a jolly well-spoken fellow & is looked upon as a kind of mascot by the WAACs.'

Beer was sending his washing home to London, suggesting that laundry facilities within the barracks were limited and that any service offered by local women (as at many other camps) could not cope with demand.

News of the Armistice was greeted at Bulford with a riot by Australians and New Zealanders, who were confronted by 300 to 400 British troops armed with sticks and iron bars. By January there was an increasingly rebellious atmosphere as men grew impatient to be demobilized. The barracks became a shambles, with soldiers desperate to get out of the Army wandering around in civilian clothing and new recruits yet to be issued with uniforms. There were many swindles, including selling Army petrol to local taxi drivers. Involved in the last was the so-called 'Monocled Mutineer', Percy Toplis, deserter and confidence man, who, in two of the many roles he assumed, walked around the barracks in Royal Air Force and sergeant-major's uniforms. He fled the area after killing one of the taxi drivers involved in the swindles and was finally trapped and shot by police in Scotland.

Increased mechanization in the 1920s led to further alterations at the barracks, vehicle standings replacing stables, for example. Since then, Bulford has been the home of many generations of British soldiers.

BUSTARD

The Great Bustard has a special association with Wiltshire, once being a native of Salisbury Plain and today appearing in the arms of the county and of the Royal School of Artillery that was established – without the 'Royal' prefix, which was granted in 1970 – at Lark Hill in 1920 and has been there ever since. Looking a bit like an unkempt turkey, it has a 6ft wingspan; the native bird became extinct locally around 1820, though migrants have since visited the county and there have been several attempts to reintroduce it as a resident.

Bustard

The bird gave its name to a remote inn on the old Salisbury–Devizes road 5 miles north-west of Amesbury. North of the inn, the road enters artillery ranges, its closure during firing being an early inconvenience for civilians who instead had to use the Tilshead–West Lavington road. Close by, the Army established a camping-ground in 1903 for summer use by Volunteer and then Territorial units. By 1910 the inn's licence had lapsed and an application was successfully made to revive it for the summer months. The *Andover Advertiser* commented that 'the establishment of an hotel at The Bustard would be a great boon to Pressmen and other civilians whose avocations call them to the Plain during the summer.'

The new licence no doubt pleased officers of Number 3 Squadron, Royal Flying Corps, serving at Lark Hill, who were quartered at the inn in the late summer of 1912 before moving to the new airfield at Netheravon.

In October 1914 the First Canadian Contingent established its headquarters at Bustard Camp, with the unit's commander, Lieutenant-General Edwin Alderson, basing himself at the inn, which had its licence extended throughout the year for the rest of the war. The inn was out of bounds to his troops, who were under canvas, as were most of his staff officers. During the terrible weather of late 1914 many developed coughs, known as the 'Bustard Whisper'. A report in *The Times* of 2 December 1914 describes the roads around the camp as being of 'almost inconceivable muddiness' and calls the camp 'Bastard', a printer's error but certainly an appropriate one. Sentries patrolling around the inn walked

Lieutenant-General Sir Horace Smith-Dorrien arrives at a damp Bustard Camp in 1912. He had become General Officer Commanding Southern Command in March.

on rifle crates sunk into the mud. Matters were not helped by the 'Storm of Bustard' on 4 December, which blew down tents and scattered not only paperwork around the Plain but sheets of corrugated iron, which disappeared into the mud to provide hazards to ankles and shins for the rest of the winter.

Canada's 1st Infantry Brigade remained under canvas at Bustard Camp throughout the winter. Oil stoves made for a very heavy atmosphere inside bell tents containing perspiring men in saturated and muddy clothing, smoking pipes and cigarettes. Bizarrely, the *Manchester Guardian* of 16 November 1914 wrote that:

> a merrier place you would not find than Bustard Camp ... After crossing so many bleak miles of downland, it is cheering to come upon the friendly fires glowing through the haze and the grey tents pressing together, as it were, for comfort.

Though many Canadians had moved into new huts by Christmas 1915, those at Bustard remained there under canvas. On the evening of Christmas Day, the 3rd Battalion had a bonfire and open-air concert, with the men enjoying a dinner financed by Toronto's city council, a gesture that was repeated on New Year's Day.

On 28 December Alderson's staff transferred to Elston House, Shrewton, but within days were marooned there by floodwater. After all the Canadians had left in February 1915, Bustard Camp was less used for the rest of the war, though practice trenches were dug nearby.

Writing of the camp in 1916 in *Direct Hit*, a journal produced by the 58th London Division, a Royal Engineer (perhaps based there in connection with the trench works) marvelled at a village of just one pub and three cottages (which were empty at the time) and at the former's signboard depicting the Bustard: 'surely no bird like it ever existed!' With the inn out of bounds, the troops had only the canteen to spend their money; for 3d one could purchase a cake of soap that would cost 2d in a town. The journal reported, perhaps facetiously, 'unprecedented excitement' when an attractive lady arrived, causing the whole company to forsake the mess.

The site continued as a camping-ground in peacetime. A 1922–23 map shows twin small structures, probably cookhouses, a water tank in Goods Clump and a bathing pond nearby.

CHISLEDON AND DRAYCOT

Very early in the Great War a camping-ground was established south of Swindon, not far from the Midland & South Western Junction Railway. The line was only single-tracked but was a useful link between the Midlands,

Salisbury Plain and Southampton. At first, the camp was named after the hamlet of Draycot, but soon became better known as Chisledon, the nearest village and railway station, perhaps to avoid confusion with two other villages called Draycott elsewhere.* The 8th Royal Welsh Fusiliers, 4th South Wales Borderers, 8th Welsh Regiment and the 8th Cheshire Regiment were the first units to be based there, under canvas, from late September.

The War Office had identified the locality as suitable for a hutted camp and so requisitioned land from T. C. P. Calley, Colonel Commandant of the 1st Life Guards, eventually buying 242 acres around South Farm for £5,725. (Calley had had a fraught journey home from Germany after war was declared, only being allowed to depart after he had pointed out he was no longer a serving officer.) Calley also owned Burderop Park, a mile to the west of Chisledon, which had been the scene of pre-war Yeomanry exercises. Probably it was this and the proximity to the railway that prompted the War Office to opt for the locality. (Most private land on which Wiltshire's wartime hutments were built was familiar to the Army from previously being used for temporary camping-sites.) But Calley convinced the authorities that the hard soil near his own house would be unsuitable for erecting huts and the site finally chosen was an unpleasant spot, 'well known as a snipe resort in the winter, and consequently a very wet one'. Initially, plans were drawn up for a cavalry barracks, but these were changed to cater mainly for infantry after it was realized that cavalry would have a limited role in the new style of warfare.

The first sign of a new camp was a long railway siding from the MSWJR to facilitate the delivery of building materials. These were then transported to where they would be used by carts willingly hired out by farmers now that the harvest was complete. By early September W. E. Chivers of Devizes was erecting a 'temporary barracks' of seventy-two structures, mainly wood-framed huts lined with asbestos and roofed with corrugated iron, at an estimated cost of £15–16,000. On the 18th, the *Swindon Advertiser* printed invitations to tender for the extension of the pre-war rifle butts near Liddington Hill, where fifty-two targets were to be added. The deadline of the 21st for submissions indicates the urgency of the situation. Swindon Town Council undertook to supply water at the rate of 1s per 1,000 gallons – 60,000 would be needed daily. On 12 October H. & C. Spackman contracted to build a further seventy-two huts.

At first, Swindon, Chisledon and Wroughton were out of bounds to soldiers, leaving them little scope for amusement. But the ban seems to have been relaxed, judging by some doggerel printed in the *Swindon Advertiser* of 6 November about the men of the Cheshire Regiment acquiring a black cat

---

*Chis*le*don was the more usual spelling during the Great War, though Chise*l*don later became the preferred form.

for a mascot in the town one Saturday. And the former Methodist Chapel in Wroughton became 'Soldiers' Rest' premises, providing weekly concerts to raise money for the Prisoners of War Fund. There was also a 'Soldiers' Rest' at Swindon Town Hall where at weekends evenings meals were served while a concert was performed.

On 1 December a gale blew down a YMCA tent and the Picture Palace, both of which had just been erected, the latter by Alfred Manners of the Empire Theatre, Swindon. When a correspondent from *The Times* visited the camp, he was not impressed, as his report published on 4 December 1914 made clear:

> Local residents describe the site as the worst that could have been selected, and doctors are against it from the point of view of health.
>
> The soil is never dry, and at present it is impossible to walk half a dozen steps in the camp field without sinking ankle deep in mud. In many places the mud is a foot deep. About a thousand civilians are engaged in erecting huts and in other work at the camp. Every one of these men leaves the camp daily with boots and socks sodden and encrusted. At first the troops were under canvas, many of them sleeping in tents which had no flooring. They lay on the ground which drank in the rain and never became dry. Huts are now being rapidly erected, and a number of the men are accommodated in these structures. Before all are up the authorities should note what the men say of the huts.
>
> They make no complaint of the daily round of duty. What they do ask – and what the British public demands – is that after the day's work the men shall be able to sleep in conditions of comfort. The new huts do not provide these conditions. Through the walls and the floor blow draughts which chill the men as they sleep. When the spring comes and the boards contract the position will be even worse. The huts, it is said, are being put up too hastily. Planks which are imperfect and from which the flawed part should be cut away are incorporated into the building. According to those who are employed in the work the felt covering is in some cases of poor quality – stuff which has been stocked too long and has deteriorated. If the fault lies with the War Office, which is paying public money for this work, that Department should be assured that the public will not grudge any sum which is necessary to secure decent conditions for its defenders. If the fault lies elsewhere it is the duty of the authorities to investigate the matter.

On the 18th the *Swindon Advertiser* added its concern:

> The condition of the camp has become a perfect quagmire. It should be renamed and called Muddeford. Decent men travel over it like a cat crossing a wet path with perpetual effort to shake off perches of land lifted at every

step ... To perfect the huts and finish the roads, men are being sent on furlough, and some are being billeted at other camps, and many of the Welshmen are gone to Cirencester ... Too many of them are on the sick list, and barking like dogs day and night ... If you ask them they tell you of the discomforts of camp life, but one seldom hears them complain.

By the time this report was printed, many of the soldiers were sleeping in barns, stables and the village hall, or had moved to billets in towns some way away. When the 10th Devonshire Regiment arrived at Chisledon on 23 March 1915, it found new huts, but still there was mud instead of roads. As elsewhere, the policy was to erect accommodation as quickly as possible and build the roads afterwards. At Chisledon reinforced concrete was used, then such an innovative technique that a hundred municipal engineers visited the camp in July 1915 to inspect the result. (Their favourable impressions were justified by the roads still existing a century later.)

Eventually there were huts for more than 5,000 men. Those using asbestos sheeting proved cold and damp. (At that time there was very little awareness of the health hazards of asbestos, especially to construction workers.) At times up to another 5,000 men were accommodated in adjacent tents and on two separate camping-sites near Wroughton. Coate Reservoir, 2 miles away, proved useful for bathing, for reasons of hygiene as well as recreation. Two privates of the 10th Argyll & Sutherland Highlanders drowned there after becoming entangled in weeds.

The downs above the camp were much used for trench-digging and exercises. During training in throwing bombs in February 1915 a lance-corporal died, and three other soldiers were injured, when his hand caught the side of the trench; the bomb fell and exploded. Three miles south-east of the hutments, the deserted hamlet of Snap, depopulated a decade or two earlier, was used as a range for artillery, machine guns or grenades, depending on which account one believes; the last two usages are more likely than the first.

By June 1915 a hospital, eventually to have 1,360 beds, had opened, and trains frequently pulled up close by with wounded soldiers from France, the MSWJR proving a convenient link with Southampton, where ships delivered sometimes thousands of battle casualties in a day. John Bates reported there in November 1915, serving until the following April as medical officer for the 15th and 16th Gloucestershire Regiment and then the 16th Hampshire Regiment – all battalions formed to train and provide drafts for the Front. Bates spent much of his time vaccinating against smallpox, inoculating against typhoid and ensuring that every man had a bath once a week. He also weeded out men who should never have been recruited in the first place; in March 1916 he reckoned 305 of the 1,000 men of the 16th Hampshire Regiment were unfit for active service. The previous December he had 'got rid' of four soldiers of the 15th Gloucestershire Regiment, one of whom could scarcely walk because of hip disease, was blind in one eye and half-blind in the other.

An ongoing problem was the pilfering of Government stores, which in 1915 appears to have been done on an organized basis, with parties of men and children taking away sackfuls of food, twelve loaves being taken to one cottage. Soldiers were complicit in the thefts, despite warnings about the practice; poorly kitted out, they themselves were not above purloining coats and mufflers for protection against the cold.

Ten huts in the camp were isolated in April 1916 so that each could accommodate thirty men suffering from 'Alien Measles'. Between March and June 1917, thirty-four men died in the hospital, nine from cerebro-spinal fever, ten from pneumonia, fourteen from measles and one from heart disease. Later that year, huts separated from the main camp by the railway were converted into an austere 1,100-bed hospital, or 'Bad Boys' Camp', to treat venereal diseases. At that time, one treatment for gonorrhoea consisted of having rubber tubes inserted into the penis to allow Condy's Fluid or potassium permanganate to enter from an overhead can; the patient then squirted out streams of red liquid. Syphilis was treated with arsenic injections in the arm, and a number of deaths occurred in various hospitals after dosages were increased and the treatment period was halved to three weeks.

Early in 1916 Chisledon became the headquarters of the new Army Cyclist Corps, with the Australian and Canadian Cycling Corps also based there.

So smart is this dining hall at Chisledon Camp that one suspects that it was for officers.

Adopted by many European armies in the days before motorized vehicles, the bicycle offered an alternative to the horse as a form of personal transport, with the advantage of not needing to be fed. On the outbreak of the Great War several cyclist battalions patrolled the coast, but as the fear of invasion receded they were reorganized into infantry, several units retraining at Chisledon in late 1915 before leaving for India in February 1916. Many men bemoaned the loss of their machines and having to march many miles carrying packs and rifles. There was still some training of recruits as Army cyclists, who would leave Chisledon to work mainly as line-of-communication troops, controlling traffic for example, and as labourers. In both cases their machines enabled them to deploy more quickly than troops on foot. But in the winter of 1918 the Corps headquarters at Chisledon closed because so many men were needed to fight as infantry. For the last year of the war third-line Territorial Force battalions (such as the 3/18th London Regiment) were based there, supplying troops to first- and second-line battalions.

After the war the camp became a demobilization centre and by the end of April 1919 had returned 400,000 soldiers to civilian life. The dangers of war did not disappear completely, for early in 1919 three boys were killed and three others injured when they trespassed on the camp ranges and picked up a live shell. In January 1921 the School of Military Administration opened at Chisledon, taking over some of the more substantial wartime buildings, several hundred others having been sold by auction. Its aim was to train all ranks in administration on business lines, with the cutting of waste much in mind.

CODFORD

A brief foretaste of what was to happen in the Great War occurred in 1872 when land north of Codford was used as a temporary camping-ground during that year's manoeuvres. But what must have been a novelty then for the village of 500 people was nothing compared to what ensued in 1914. On 11 September the *Warminster Journal* advised that 24,000 troops of Kitchener's New Army were to camp in the neighbourhood. Units such as the 8th and 9th Loyal North Lancashire Regiment and 10th and 11th Cheshire Regiment reported there to form part of the 25th Division – and lived in tents pitched by Territorials of the 4th Wiltshire Regiment until huts could be built. Many of the arrivals were in civilian clothing, a few had been issued with a mixture of obsolete blue and red jackets. Some NCOs (the senior ones being 'old sweats', the junior ones often arbitrarily selected from the recruits) wore bowler hats as a sign of authority.

Eight hundred men of the 10th Cheshire Regiment had no change of clothing and were given a blanket apiece, their provisional NCOs being chosen by what they had done in civilian life. The same regiment's 11th Battalion on arrival:

> turned into a field and found itself without tents, blankets or food. The Colonel and two or three officers went into Salisbury and bought what they could, and presently rations were being drawn and a tent pitched for every twenty men. The men had been told to bring their ordinary clothes. They sweated in these and had no change. To wash, they went down to the river, stripped, dipped and ran about to dry.

By 25 September instructors from the Duke of Cornwall's Light Infantry, the Grenadiers and Royal Marines had arrived – and perhaps these professional soldiers were horrified at the task facing them! On 20 October the 7th King's Shropshire Light Infantry received temporary 'Kitchener blue' uniforms.

The River Wylye was used for the baptism of at least three soldiers. Less welcome to the authorities were the pacifist preachings of a member of Spurgeon's Tabernacle (a narrow Calvinist denomination founded by Charles Spurgeon in the previous century), who was arrested for sedition.

The King's Shropshire Light Infantry's regimental history recalled that at Codford:

> From October 15 until November 10 it rained in torrents very day. Roads to the camp became impossible, and training was suspended. Even route marches were impossible ... the men, of course, had no change of clothing, and no washing accommodation. There was nothing to be done day after day but to lie, in an indescribable state of mud, in tents without floor boards, listening to the rain beating on the canvas.

It is said that other units held mass meetings and some refused to parade. After weeks of sleeping in the mud, still with one blanket each and with a wet winter setting in, men of the 11th Cheshire were among the unhappy and 'serious trouble was narrowly averted' by an issue of beer and a move to billets in Bournemouth. One exaggerated story has it that a particularly discontented Welsh battalion had to be disbanded or 'marched away'. The 10th Welsh Regiment (1st Rhondda) did move on 30 September to join other Welsh units, but only after politicians had expressed concern about Welsh-speaking battalions being trained alongside English units. But the battalion's two or three weeks at Codford in fine weather were hardly long enough for patriotic fervour to turn into serious discontent. However, after two months of appalling weather, the 10th Devonshire Regiment's B Company (consisting of nearly all Welsh miners who had been working in Devon at the start of the war) rioted but were subdued by a dressing-down from their colonel.

Codford

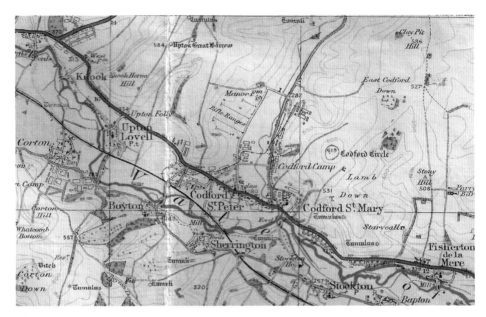

An extract from a 1in Ordnance Survey map printed in 1920, showing Codford Camp and its military railway. The huts at Corton are also shown, but by the time of the survey a smaller group west of Boyton Church had been removed.

At first, the George Hotel housed part of the divisional headquarters, hardly encouraging off-duty relaxation there for the rank and file. The bar was open to troops for an hour at midday and from six to nine in the evening, with an overflow marquee being hastily erected. At Upton Lovel a mile away, an old shed at the Prince Leopold Inn was converted as living quarters for soldiers and workmen. (Later a military power station comprising several buildings, including engineers' living quarters, was built in the village.)

The YMCA provided a marquee in the centre of Codford that could hold something like 1,000 men; by early 1915 a YMCA refreshment and recreation hut had been built. The Congregational School also offered recreation, books, tea and coffee to off-duty soldiers. Another early facility was Albany Ward's Picture Palace, one of two cinemas to be opened in the village, with seating for 500. A Red Cross hospital was set up at Codford St Mary Church, with more serious cases being treated in Salisbury until the camp hospital was ready for use. Cases of enteritis – or aacute intestinal problems – resulted from some of the men eating horse chestnuts.

When Norton Hughes-Hallett arrived with the 7th King's Shropshire Light Infantry in early October 1914, he wrote home expressing the hope that chairs would soon replace boxes in the officers' mess – though his battalion had plenty of blankets, and discussions were going on with Harrods about catering. In the meantime there was often trouble at the railway station as ration parties argued over incoming stores. Warminster was out of bounds because of scarlet fever. His men were sleeping sixteen to a bell tent, with no bed-boards. A new commanding officer arrived, a former Indian Army man who had been retired and had worked as a golf-club secretary for some seven years; he was so shattered by what he found that, like Hughes-Hallett's first CO at Tidworth (Colonel W. E. Sykes of the 9th Worcestershire Regiment), he shot himself.

Until huts were erected, many buildings in Codford were taken over and used as offices by the Army. By 15 September 1914 the London City & Midland and the Capital & County banks had branches in the village. Temporary shops, cafés and barber's saloons sprang up on any vacant site and even in gardens, giving a shanty-town appearance to the hitherto quiet little village. The Rev L. L. Jeeves, chaplain to the 55th Infantry Brigade, described it thus in the summer of 1915:

> The old sleepy village is half filled with horrid booths and shanties, where tobacco, hosiery, and a thousand odds and ends can be bought at an increased cost, for the owner of the property has asked £1 a week in rent for a glorified cupboard which now constitutes a shop. A cottage lets its front parlour to a bank.

In early October 1914 Sir John Jackson Ltd started to build fifteen hutted camps at Codford, with crowded workers' trains leaving Salisbury at 6am and 6.10am, the men being issued with tokens to authorize them to travel. By the end of the month the first huts were nearing completion at Fisherton de la Mere, but early in November 33,000 of the soldiers camped out in the area moved into billets, with 2,000 remaining to help with camp construction. When the huts were ready, the 10th Essex Regiment found that:

> John Jackson's mansions were not palatial, nor the pleasantest abodes when a full summer's sun beat down upon their oven-like roofs and sides ... there were compensations in the shape of shower-baths and a running stream at the bottom of the garden which helped towards contentment.
>
> Boots, boots, boots, went slogging up and down day in day out over Wiltshire until the 18th Division could really lay claim to be able to walk a little bit. Then there was trench-digging at Yarnboro Castle [Yarnbury, the earthwork 4 miles from camp], and manoeuvres over Stony Hill [between the castle and Codford Camp], bomb-throwing with Heath-Robinson jam-pot contrivances and gas-mask drill with cotton-waste and black crape, Lewis-gun classes in the nullah [stream or ravine] behind the camp and range-finding on the hills above, early morning runs up precipitous slopes, which nearly killed ... antediluvian members of our cosmopolitan unit, and night-marching lectures conducted by an officer eager to sell his own book on the subject.

In March 1915 two privates of the 3/8th Manchester Regiment forced their way through a stained-glass window into Codford St Peter Church, breaking the altar Cross, tearing leaves from the 300-year-old Bible and setting fire to church linen, vestments and altar cloths stored in a chest. The two culprits pleaded being under the influence of drink.

In July 1916 Codford became a New Zealand Command Depot for men who were convalescing from wounds or sickness and were not yet fit for active service, with Australian training battalions also based there. The camp hospital became Number 3 New Zealand General Hospital and eventually contained beds for ten officers and 980 soldiers. At the end of 1917 there were 400 patients, with another 200 convalescing.* Codford was also the centre for the New Zealand forces' dental service in the United Kingdom.

On joining the camp, a convalescent was graded B3 and given light work, progressing to B2 and route marches of 4–6 miles as he grew stronger. At B1 level the march was extended to 8–10 miles a day. An 'A' classification meant a man could march 14 miles a day and was fit to join a reserve unit. (This is a simple set of gradings and there were more complex systems, an Australian handbook listing thirteen different ratings, including 'B1A2: Fit for overseas training camp in three to four weeks' and 'C2: Unfit for overseas service, temporarily unfit for Home Service'.)

A popular facility was the Aotearoa Club ('Aotea Roa' being the Maori name for New Zealand), run by ten ladies and regarded as the best-equipped club on the Plain, with reading, writing and games rooms and library. A 'soldiers' recreation hut' was opened by the Catholic Women's League in November 1916 and was available to all regardless of denomination; it was used as a Catholic church on Sundays.

The Australians' Number 4 Command Depot moved from Wareham in June 1917, staying until November, when it left for Hurdcott. As was the case with the New Zealand depot, its purpose was to rehabilitate ill soldiers towards fighting fitness. That year, 45 acres of land near the camp were planted with vegetables as part of the nationwide efforts to boost food production; one plot measuring 42 by 185ft yielded 3 tons of beans. Beekeeping was another pursuit the soldiers were encouraged to follow. (Perhaps the bailiff of Manor Farm saw some irony in this, having frequently complained about how military activities were inhibiting his own efforts to produce food.) A rather more energetic pastime that year was undertaken by Australians who cut a badge 175ft long and 150ft high on nearby Marmpit Hill – called 'Misery Hill' by the soldiers who in training had to run up and down it with packs on their backs. As with the Australian emblem near Fovant, that at Codford is popularly known as the

---

*Different sources give varying numbers of beds at all of Wiltshire's camp hospitals, as numbers fluctuated through the war.

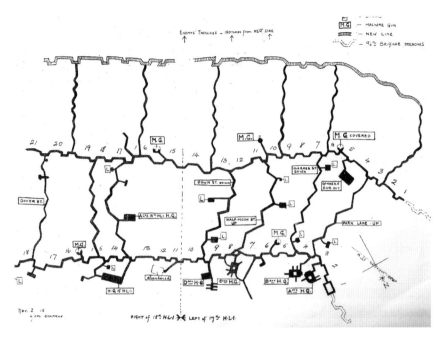

This contemporary plan shows the network of trenches close to Yarnbury Castle used by the 96th Brigade (including the Highland Light Infantry) in late 1915.

'Rising Sun'. (One story has it that the badge was cut by Australians awaiting demobilization in 1919, who are said to have used a mixture of sand and coloured glass in the construction. However, a near-contemporary postcard states the badge was made in 1917 by the 13th Battalion, Australian Imperial Force.)

On the day that the Armistice was announced, all was quiet at Codford until night fell. Rations had not been delivered that day and more than 200 drunk and hungry New Zealanders clamoured outside Number 14 Camp's cookhouse demanding to be fed. The cooks (ten of the twelve being boxers) armed themselves with ladles of boiling water and four side-arms. The men slunk away, but, reinforced by up to 500 comrades, marched on the camp cells to demand the release of seven prisoners. Military policemen, the camp commandant and adjutant were overwhelmed and substituted for the inmates, but were quickly released by other officers, who wisely did not attempt to round up the original prisoners.

On 28 November 1918 there was a riot at the George Hotel involving New Zealand and American troops. The hotel manager – who had not experienced any problems in the past three years – had been tipped off that there was likely to be organized trouble. An argument developed over payment for drinks, a soldier threw a mug through a window and tills, drink and tobacco were stolen. Military policemen prevented the trouble escalating. Of six accused New Zealanders (all of whom had seen active service in France and five with 'bad' military records), four received six months' imprisonment with hard labour, one was acquitted and one bound over.

With the return of peace, the railway built to facilitate the construction of the camp was used to dismantle many of the buildings (though some were retained for domestic and farm use). Much surplus equipment, including ordnance stores, mess equipment, tents and food machines, was sold by Woolley & Wallis in the autumn of 1919. A 1926 map marks only disused rifle butts and a very few huts scattered around fields.

An ANZAC cemetery at the eastern end of the village contains the graves of thirty-one Australians and sixty-six New Zealanders.

CORTON

See 'Boyton and Corton'

DEVIZES

Le Marchant Barracks opened in 1878 at a cost of £46,000 and were the headquarters of the Wiltshire Regiment. They were named after Sir Gaspard Le Marchant, who thirty-six years before had commanded the 99th Regiment of Foot, destined to become part of the Wiltshire Regiment.

The barracks had accommodation for 250 men with married quarters, cricket ground and 'recreation establishment', which included reading and writing rooms, bar and billiards room. A hospital had two thirteen-bed wards with a separate two-bed infection bay.

With the outbreak of the Great War, the barracks were extremely busy with the mobilization of Reservists and the formation of new battalions. Leon H. Todd recalled: 'When we arrived there we found the depot overcrowded & no one knew where to get a place to sleep or to get something to eat, & for two nights I slept out on the barrack yard, so my first start was not a pleasant one.'

The *Devizes Gazette* of 27 August 1914 noted that there was 'a rather tight fit as regards accommodation', and:

> on the greensward of the gymnasium [open-air area used for 'Swedish drill' or physical training] there is a canvas colony. Wadworth Ltd [the Devizes

Devizes

Everyone and everything is spick and span inside this room at Devizes Barracks in 1911.

> brewery] have lent the authorities their big marquees, besides which there are numerous bell tents … there is plenty of food for all … there has been a scarcity of knives and forks … New clothing and equipment are coming to hand; of the former there is almost, if not quite, enough to fit out the new unit, and a large proportion will soon have equipment.

It reported that 800 recruits had enrolled so far, mostly in the new 5th Battalion; of these, 75 per cent were Wiltshire men, but this proportion was declining. Christopher Hughes was later to note that the 7th Wiltshire, also newly formed, comprised men from Wiltshire, Birmingham, South Wales and Wolverton in Buckinghamshire. Men from Wiltshire also found themselves in regiments affiliated to other areas. Hughes wrote:

> I have often wondered if the War Office worked on any system in the drafting of men to the various County regiments, why a large number of the first recruits from Wiltshire should have been drafted to Irish regiments, it is hard to say unless it was to delude the public into thinking that the Irish had changed their attitude towards this Country and were anxious to help us.

On 2 September Hughes had cycled the 13 miles from Marlborough to Devizes to see if he could enlist in the Wiltshire Regiment, only to find that:

> recruitment was for the moment stopped, the number of men reporting had so greatly exceeded the accommodation at the Barracks that for some nights men

had been sleeping under the hedges without food and without money to buy food, and some had therefore gone home, many others had been sent home while those waiting at the depot were absorbed into units.

Huts were soon erected to ease the problem, but the barracks continued under pressure, having to cope with the 3,000 Reservists and 5,000 recruits who reported there in August and September. Field Marshal Lord Methuen, a Wiltshire resident, wrote to Lord Kitchener about the lack of arms, uniforms, blankets, drinking cups, plates and dishes. 'At once one of Harrods' head men came down to me from Lord Kitchener giving me a *carte blanche* to get what I required for the depôt. Now the complaints are vanishing,' Methuen reported.

Early in the war part of Devizes Prison was converted into an Army detention barracks, able to accommodate 160 men. During the winter of 1914–15, Canadians from nearby tented camps were prominent in the town, making much use of bathrooms at 20 Northgate Street, and their divisional artillery moved its headquarters to the Bear Hotel in Devizes. Locks on the nearby Kennet and Avon Canal were used to train troops who were to operate craft on the Belgian and French waterways.

DEVIZES WIRELESS STATION

Wiltshire had a key role in the early development of wireless transmissions for military purposes. If the claim that at Lark Hill on 27 September 1910 the actor Robert Loraine made the first transmission to the ground from an aircraft is disputed (perhaps even with the qualification that it was the first to a 'military installation'), the activities of the 'Father of Wireless', Guglielmo Marconi, are well chronicled. In 1896 he experimented at Bracknell Croft on Three Mile Hill, 3 miles north-east of Salisbury, his equipment including copperplate aerials that were 2ft square and 25ft from the ground and much larger plates 10ft from the ground, with aerial wires 90ft long on bamboo supports. From there, he transmitted up to 1.75 miles, assisted by the General Post Office and Royal Engineers and in front of Army and Navy representatives. (A painting of this event by Steven Spurrier includes Royal Navy officers and red-coated soldiers.) Marconi returned to Bracknell Croft in March 1897 and transmitted to various points in the locality, this time using aerials on kites. 'Indications of signals' were detected at Beacon Hill, 6.8 miles away, with 'good and reliable signals' being received at a point 5 miles distant, between East and West Grimstead. Marconi returned again in 1900, when he communicated with Bath, 34 miles away. On at least the earliest visits to Bracknell Croft, he lodged nearby at Hillcrest Bungalow, where brass tubing found in the 1970s was presumed to be a relic of the experiments.

In August 1913 Parliament approved a plan by Marconi's Wireless Telegraphy Company to build an imperial chain of six wireless stations linking England to Australia. The receiving station, consisting of eight tubular steel masts (six of which were 300ft high with two of 50ft at either end), was built on Morgan's Hill, above the village of Bishops Cannings and 3 miles north of Devizes. In December the War Office (which after an early lack of enthusiasm was now very interested in the military potential of Marconi's work) was concerned about lack of security at the two English stations in the chain – the receiver near Devizes and a transmitter at Leafield in Oxfordshire, which were administered by the General Post Office. It asked for fortifications and bulletproof shelters, and for contingency plans to be made by the GPO and police for protection 'against ill-disposed persons'. The Marconi Company indignantly questioned the necessity, though in May 1914 women believed to be suffragettes entered the Leafield site, prompting the appointment of a nightwatchman. In the end, the Postmaster General agreed to erect interlocked wire fences around the masts and buildings, but in June 1914 the Government decided 'not to take any measures for the defence of the receiving station at Devizes in time of war', though the station at Leafield would be surrounded by a fence.

Shortly after the outbreak of hostilities and with the Devizes station all but complete, the Government 'repudiated' its contract with Marconi's (leading to the company claiming £7,181,774 in damages and being awarded £590,000 in 1919). In the first months of the war, experiments 'near Devizes' and presumably at the wireless station found it was possible to determine the point of origin of a wireless transmission.

This led to the Royal Engineers moving on to the site to adapt the existing buildings and equipment. In August 1915 the *Wiltshire Advertiser* reported some 150 men of the 499th Company, Motor Transport Section, Army Service Corps arriving at Shepherds Shore, which was the location of one of the terminal masts, apparently for a short stay under canvas, with the townsfolk attending a gala sports day a week or two later. It is likely they were there in conjunction with the conversion of the station. In November the Royal Engineers drew a plan showing a 1,000ft circle near Shepherds Shore, possibly indicating a new installation of four masts in a circle designed to help determine from which direction a signal was coming.

An Army intelligence unit from Langley, near Slough, moved to the site, which became one of three direction-finding stations in Britain charged with intercepting signals from Zeppelins. By 1918 it had become the headquarters

Devizes Wireless Station

The wireless station near Bishops Cannings in the 1920s. One hut still stood ninety years later.

for all Army wireless stations in England and was also a training centre for wireless operators involved in signals intelligence. That year, Ernest Gill was at the station as its commanding officer. He was a major in the Royal Engineers and also a member of a branch of military intelligence, MI1e, responsible for breaking codes and analysing signals traffic. At the time of the Armistice, semi-permanent buildings were being erected to the extent that the establishment's size was being doubled. When Gill suggested this might no longer be necessary, he was informed: 'It is not for junior officers, knowing nothing of the facts in the knowledge of higher officers, to put forward suggestions of policy.' So construction went on for more than twelve months and on the day it was completed, demolition (which must have been only partial) commenced. Gill recalled that during those twelve months 'instead of taking note only of the German wireless messages and locations, we had also to keep an eye on all the messages sent by our late Allies and our work was more than doubled'. But, until he protested, members of his staff were being demobilized, making the extra duties difficult to fulfil.

In 1920 the station was taken over by the General Post Office and its equipment used to communicate with vessels at sea; an Ordnance Survey map revised in 1922 shows a cluster of a dozen buildings close to one of the central masts. After a new station was built at Portishead near Bristol the masts were dismantled in 1929. Gill visited the site around 1931 and found only overgrown foundations, though one building and the concrete foundations of the masts could still be seen in 2011 (such as at map reference 032664).

DURRINGTON

Durrington Field, a mile west of Durrington village, was used as a camping-site in the manoeuvres of 1872 and later was one of the first permanent camping-grounds on Salisbury Plain. A Royal Field Artillery 'practice camp' (map reference 134444) to the east of the more prominent Lark Hill Camp was in operation there by 1903. The site was among those built on during the massive development of hutments in the locality at the start of the Great War.

A report in 1914 had recommended that the School of Gunnery move from Shoeburyness to Salisbury Plain and of six sites considered the one chosen was at Durrington Walls, an ancient earthwork to the east of Durrington Camp. There would have been student accommodation for forty officers and forty-five NCOs, together with buildings for staff and equipment. Though it was noted that the site of an ancient 'British Village' was close by, it was not thought that this would be at risk. The outbreak of war meant the plans were postponed and the far more extensive building of hutments took place to the west, at Lark Hill and on the Durrington camping-site itself.

The School of Gunnery did move in 1915, to Lark Hill's Number 14 camp, with the understanding that it might eventually establish itself at Durrington Walls. Luckily this never happened, for in 1925 the important archaeological monument of Woodhenge was discovered close to the proposed site. As it was, the wartime huts of Durrington Camp came to within feet of Durrington Walls.*

In the House of Commons on 28 November 1917, the Under-Secretary of State for War, James Macpherson, was asked if an inquiry would be made into:

> complaints about the conduct of the second lieutenant, E. L. Smith, commanding the 3rd Southern Non-Combatant Corps at Durrington Camp, Larkhill, who, since he took charge of this company, has persistently carried out a policy of irritation, his latest act being to order the searching of kitbags of the men and their personal belongings and their private correspondence; and among other orders he issues, which are not imposed in any other company, are the maintenance of the summer practice of washing and wet-scrubbing huts, the consequence of which is that their huts often remain in a damp state for two or three days; and also his practice of making all available men parade at 6.10 a.m. for route march on empty stomachs, this being the only company on Salisbury Plain where this practice is followed?

---

*As do present-day military buildings. Woodhenge has similarities to Stonehenge, the designs of both appearing to be influenced by where the sun rises on Midsummer Day. Stone markers now indicate the positions of the original wooden posts.

Mr Macpherson replied that a full report had been called for, but its findings appear not to have survived. (The Non-Combatant Corps was formed in 1916 to accommodate conscientious objectors who were prepared to carry out uniformed military service of a non-combatant type, such as unloading stores and maintaining camps. Its officers were medically downgraded men, many of whom had seen active service.)

On 2 July 1919, four days after the Treaty of Versailles formally concluding the war had been signed, men of the 3rd West Yorkshire Regiment based at the camp refused to parade with fighting equipment, leading to seventeen with the longest service being court-martialled. They argued that they had joined only for the duration of the war and, as 'Peace had been signed', they were not going to do any more training; they had also resented the company sergeant-major calling them 'a lot of bastards', though he claimed he had said 'a lot of Bolshevists'. The accused were sentenced to between twelve and twenty-one months' hard labour, remitted to nine months. One was acquitted of a charge of leading 200 men in an attempt to release prisoners from the guardroom, and a second from a charge of yelling for it to be burned down, the latter's defence being that having been gassed he could not shout.

Though throughout the war there were references to Durrington Camp, it was regarded very much as part of the Lark Hill complex, their respective hutments being contiguous. Some maps wrongly show the western block of buildings as being Durrington Camp and the eastern as being Lark Hill. As the war progressed, 'Durrington Camp' became less used, probably to avoid confusion with the town of the same name in Sussex, and it lost all identity in the 1920s.

FARGO

Fargo, or Fargo Down, camping-ground (map reference 109441) was established in 1904 north of Fargo Plantation near Lark Hill and mainly accommodated artillery units, with the Honourable Artillery Company being a regular visitor.

The site was also used by medical units. *The Times* in August 1917 reported how three ambulances had carried out operations near their camp at Fargo, with Territorials acting as wounded and receiving first aid:

> The Territorial medical officers treated the supposititious cases with temporary dressings and tourniquets on the spot, after which they were removed to dressing stations, there to await operations if considered necessary. When it is

Introduced in 1904, this 60-pounder gun, seen here at Fargo, became the mainstay of the British Army's heavy weaponry and could propel a shell 10,300yd in under 15sec.

not possible to carry out practical work in the field the ambulances are busy with theoretical instruction in subjects of wide scope, including that of sanitation and hygiene as effecting the health of troops in camp.

In July 1914 men of the Royal Army Medical Corps attached to the Wessex Division camped there, with the locality's medical tradition continuing later in the year when work started on the 1,200-bed Fargo Military Hospital, built on the site of an isolation hospital for horses. Australian patients were unimpressed with the hospital food and corrupted the name to 'Starve-oh'.

Some wards at Fargo were for prisoners of war. The Australian James B. Barclay, who was nearby at Rollestone Camp, wrote home that 'over from us there is a German hospital, with 35 patients, and 19 legs amongst the lot. The German prisoners get the best of treatment, which they do not give to our boys'.

During the war inquests into fatalities at nearby camps and airfields were held at the hospital. A local newspaper in December 1917 referred to a doctor of the United States Army Medical Corps being based there. He was probably one of those American military doctors who arrived from September onwards to replace British civilians in military hospitals; the War Office explained that it was preferable to employ full-time doctors who did not have to respond to the 'exigencies of their private work', and that in any case there was an urgent demand for orthopaedic surgeons, which the United States could meet.

After the war the hospital became disused, though some accommodation was retained as married quarters for RAF officers based at Rollestone. Given its size, it is remarkable that hardly any photographs of it have been traced and there are very few references to it in memoirs and diaries. There are a few allusions to 'Fargo Hospital Cemetery, Durrington', which was the cemetery then belonging to the parish council off the Amesbury–Upavon road. Private Andrew Maclaren of the 10th Company, Army Service Corps 'died of blood loss consequential to a stomach ulcer and exhaustion 1st March 1918 aged 46 years at Fargo Military Hospital' and was buried in the 'hospital cemetery', as was his comrade from the same unit, Private E. Fort, on 4 March. Both men are missing from the Imperial War Graves Commission register of graves in Wiltshire, so it may be that their families arranged for their bodies to be moved to near their homes, with the Commission not being involved with the reinterments. Most of the 141 Australians and thirteen Americans buried in Durrington Cemetery would have been patients at Fargo, the latter's bodies subsequently being removed, as were those of forty-six German prisoners of war and a German and Austrian internee.

FOVANT

In both 1872 and 1898 Fovant, 8 miles west of Salisbury, had a taste of what was to befall it in the next century when troops camped at West Farm during manoeuvres. Perhaps the Army noted the potential of the site, for soon after the outbreak of the Great War a vast spread 'of green corrugated iron huts with verandahs in front of many of them' was built near the village to provide a training and transit camp. It was close to the London & South Western Railway and Dinton Station and within easy reach of Southampton, a major embarkation port for France. There was good open downland to the south for training and though the camp became muddy it was not so liable to flooding as those in the Wylye Valley. Several contemporary comments suggest that its huts were more comfortable too.

Of the numbered groups of huts in the area, each providing quarters for a battalion, Numbers 1 to 4 and 13 were close to Fovant itself, others being at Hurdcott, 3 miles to the east, and at Sutton Mandeville, a mile to the west. There was sizeable accommodation for the Army Service Corps, the Royal Engineers and the Royal Field Artillery and a hospital whose number of beds increased from 150 to 601. A 'Pictures and Varieties Palace' was built and, after some adaptations, became the Garrison Cinema. On its demolition after the war, its steps were retained in a house built on the spot (map reference 010290).

When traction engines had trouble hauling supplies up the steep hills and turned the roads into quagmires, it was belatedly decided to lay a railway track from Dinton, though elsewhere in Wiltshire one of the first steps in

An early view of the cinema at Fovant Camp. In its life of only some five years its exterior was altered and it was renamed the Garrison Cinema.

constructing a new camp was to lay down sidings from a main line to help the delivery of building materials.

British regiments were based at Fovant from March 1915 with one of the first to arrive, the Oxfordshire and Buckinghamshire Light Infantry, finding the huts 'well built, waterproof, and new, with plenty of space round them, and with large areas for parade grounds', though the rifle range had not yet been 'properly finished'. As it was, 233 men of D Company of the 8th Battalion were put to work building ranges, sheds, huts, cisterns, roads and wells. In April a soldier reported that 'I think we shall be quite comfortable here, we are in wood houses & we have very good beds, all new & clean.'

The camp in late 1915 is depicted in forty-seven photographs out of 175 published in *Snap Shots of the 15th Battalion The Prince of Wales's Own (West Yorkshire Regiment)*. The 15th arrived at Camp Number 1 in early September, when large buildings at the entrance were still being built and the unsurfaced roads marked out by white-painted stones. The photographs feature training on Lewis machine guns, trench mortars and trench catapults (some of which were improvised larger versions of the schoolboy 'Y' catapult, intended to extend the range of a hand grenade but hazardous to the user).

In contrast to these favourable impressions, when the London Rifle Brigade arrived in January 1916 it found that still hardly any of the roads in its hutments had been completed and the drainage from other hutments ran on to its ground; a lack of duckboards added to the unpleasantness. The spring weather was very wet and the parade ground – on a steep slope – so slippery that the men could hardly stand on it. The regimental history notes that the recreation facilities were 'not great' and an attempt was made to lay down two concrete practice wickets. It also observes that there was already a good set of trenches, so facilities to practise digging were not numerous. Perhaps this lack of 'facilities' led to the idea of familiarizing the men in pick-and-shovel work by having them construct the military badges on the nearby hillside.

This was the reason given for the making of two tennis courts near the officers' mess, which proved a boon to officers and sergeants – and have been said to have led to a question in Parliament about a hundred men being employed in making tennis courts for officers! No such reference has been found in *Hansard*, the official report of Parliamentary proceedings, but in May 1918 Daniel Combes of Manor Farm, Dinton, a member of the Wiltshire War Agricultural Committee, did complain that two men had spent eight days at a camp near him (which suggests Fovant) scything a lawn so officers could play tennis. The official explanation was that the men were doing the scything in their own time, which was why they had taken eight days, that the lawn had been provided by a local resident and that tennis enabled officers on light duties to become fit for overseas service. (From 1916 the War Office had placed increasing importance on the role of sport in keeping servicemen healthy and fit. A cinder tennis court was laid down at nearby Hurdcott in 1917 for convalescing Australians. Much sports equipment of all types was included in the surplus goods from local camps auctioned off after the war.)

As well as the YMCA having a hut in the camp, it provided its usual facilities at Manor and West farms and ran a guest house where friends and relatives of hospital patients could stay free of charge.

One member of the London Rifle Brigade was W. H. A. Groom, who trained at Fovant in the summer of 1916 and was one of the small band of soldiers who enjoyed his time in Wiltshire: 'What nostalgic memories Fovant conjures up,' he recalled in *Poor Bloody Infantry*. 'The six months at Fovant were the best part of my army life. For one thing the camp was situated in a lovely quiet part of Wiltshire and on one side of the camp was a high, smooth grass-covered escarpment.' But Groom found bayonet training to be 'the biggest time-waster . . . as not one hundredth of one percent of the infantry in the 1914/18 war ever used a bayonet'.

Bernard Livermore described his experiences as a member of the Queen Victoria's Rifles at Fovant in *Long 'Un – A Damn Bad Soldier*:

> Fovant was our next stay. In the summer a delightful little village near Salisbury. In the winter the mud surrounding our huts was always ankle deep.

Night Ops and Forced Marches were constantly appearing on our Battalion Orders, there were always tiring fatigues to be performed ...

During the day we were marched up and down, sometimes at the double. We handled our Mills Bombs with considerable apprehension, fearing that they might explode before we drew out the pin and lobbed them out of the trench. A few months later we carried a couple in our pockets and soon lost our fear of these handy little weapons.

Australians were prominent at Fovant from August 1916, becoming the major occupants the following March. 'This camp is considered to be one of the best in England, it lies among hills, and is well sheltered. In this respect it is a great improvement on Rollestone, which was greatly exposed,' said Australian Hugh Welsh. August 1916 also saw the Royal Engineers quarters near West Farm being turned into a compound for German prisoners of war who worked on local farms.

With the return of peace, Fovant became a major demobilization centre, handling from 2,000–3,000 British soldiers a day. One Australian soldier died at the camp hospital as late as 2 February 1920; he was William Symons, who had stayed on in Britain after the Armistice to study in Edinburgh. He had contracted tubercular disease and been admitted to the hospital on 19 December 1919.

The camp closed in November 1920 and the buildings were demolished. Much of the land on which they stood had been sold the previous year, when the Earl of Pembroke's estate at Fovant (which included much of the village) had been auctioned.

In 1924 a cross was erected in Fovant churchyard in memory of forty-three Australian and twenty British soldiers buried there after dying in Fovant Military Hospital.

HAMILTON

A camping-ground (map reference 108448) was laid out to the west of Lark Hill Camp in the summer of 1909 and named after Sir Ian Hamilton, the departing General Officer Commanding the Salisbury Plain District. It was in use by August and, judging from the amount of mail sent from the site in 1910, had a large concentration of troops camped on it during that year's manoeuvres.

Troops were under canvas on the site early in the Great War while thirty-four groups of huts were being hastily erected on either side, at Rollestone Bake and Lark Hill. Even more primitive shacks were built at Hamilton itself.

Hamilton

Taken in the winter of 1914–15, this photograph shows the newly laid Lark Hill Military Railway (on which men are working in the mid-distance) and the primitive Aylwin huts, which their occupants thought were more suited to animals.

The 10th King's Royal Rifle Corps was quartered there in April 1915, in 'a species of rabbit hutches, a kind of dug-out above ground, with canvas nailed over the frames, and going by the name of Alwyn [*sic*] Huts'.* When a celluloid window in one caught fire, a whole line of huts burned down and the 10th moved to Camp 13 at Lark Hill. The remaining 'hutches' were among the first to be cleared away with the return of peace.

Men of the 11th South Wales Borderers spent a short time on a firing course 'at Lark Hill' and were put in 'quarters appeared like chicken pens. The lads quickly responded ... by poking out their heads and cock-a-doodling'. One guesses these were the shacks at Hamilton. There was unrest when A and B companies' arrival was delayed until after midnight and they found no rations were available. Nor was there any breakfast, and when A and B were ordered to fall in at the butts, they refused. A sergeant, corporal and six

*Francis Aylwin, formerly of the North West Mounted Police, had designed huts consisting of canvas stretched on wooden frames with mica windows and wooden floors. They could house six men and it was claimed that they could be erected in under two and a half minutes. Thousands were ordered for Salisbury Plain and other military centres. Aylwin had made himself at home in such a building in sub-zero conditions in British Colombia, but claims that they provided weatherproof and comfortable accommodation for soldiers proved unfounded and they were soon discontinued – though in 1919 they comprised part of the Contagious Compound at Sutton Veny Hospital where they were 'worn out'.

privates were arrested and appeared before a district court martial. The NCOs were stripped of rank and all got two years' detention.

Hamilton lost its identity as a camp soon afterwards, though there are references late in the war to 'Hamilton Lines'. Hamilton Battery was established on the eastern end of the site and in the early 1920s consisted of two 9.2in howitzers on semi-permanent emplacements and 8in and 6in guns on wheeled carriages. It was rather close to living accommodation at Lark Hill and firing practice there was a very noisy feature of camp life. The battery was served by a short spur off the Lark Hill Military Railway. This was linked with the battery by a 750yd tramway and there were a number of small huts and one larger building alongside, presumably for the storage of ammunition.

HEYTESBURY

Fifty acres at Knook were selected for military use in September 1914 and an artillery camp was laid out there later that year, adopting the name of the nearest railway station, with Heytesbury House being taken over for officers' quarters. Belgian refugees were among the construction workers.

A Squadron of the 1st Lothians & Border Horse Yeomanry arrived on 30 July 1915 to find that the camp had recently been vacated by a Royal Field Artillery unit attached to the 18th Division. 'The camp was in a dirty condition and stables were very wet after a great deal of rain,' noted the Yeomanry's war diary. 'We had no rifles however, as they had been left at Dunbar for the 2nd Regiment.' On 6 August the squadron received thirteen remounts, 'cobs at 15 hands' and probably from America, some of which appeared to be 'quite useful animals'.

Harry Siepmann of the 169th Brigade, Royal Field Artillery, was based at Heytesbury in 1915 and recalled in *Echo of the Guns* that 'the dismal prospect of Christmas on Salisbury Plain loomed before us until finally it became depressing reality'. Many of the men were from Lancashire and lacked the money to get home, so mutiny and desertion were predicted. Three days before Christmas, a cordon of guards and military policemen (some of whom were themselves thought likely to desert) was placed around the camp and at the railway station. In the event, nothing happened, though Lady Heytesbury's agents complained of men 'breaking off branches and stealing green stuff' to decorate their huts.

Dennis Wheatley reported to the camp in August 1916 as an RFA subaltern. He found Heytesbury House was 'set in a pleasant park [and] a pleasant Georgian house which gave us ample accommodation for our mess on the ground floor and, perhaps, twenty bedrooms'. Huts were also erected in the garden. Ever fond of the good life, Wheatley thought the location isolated and bereft of amenities. (However, when billeted in the even more isolated Tilshead, he did enjoy a relationship with the wife of a local racehorse trainer.)

Young officers of the RFA's 291st Brigade, 'used to the glamour and gaiety of London', also found Heytesbury dull. Writing in *Direct Hit*, the journal of the 58th London Division, one noted that the few women he had seen there 'seemed to occupy the attention of a few of our privileged leaders, and I believe they receive kindly attention too'. The only other attraction, apart from inns, was the local 'Picture House', which offered films and 'occasionally … a splash of local "stars"'.

From 1916 Australians occupied the camp. The site was retained by the Army after the war, though Heytesbury House reverted to civilian occupation and in 1933 was bought by Siegfried Sassoon, the celebrated war poet. In later years the camp changed its name to Knook, an early usage being in a list of Salvation Army huts compiled in 1919 (when it was rendered as 'Knock').

HURDCOTT

Built early in the Great War near a small hamlet west of Salisbury, the camp at Hurdcott was closely linked administratively with that at Fovant. In January 1916 the 4th London Regiment found 'the Hurdcott camps were arranged on suitably designed principles with well ventilated sleeping huts and roomy messing and recreation rooms'. There were excellent training and sports grounds spread over some 5 acres. The 4th constructed a bayonet-fighting assault course close to a bombing ground and improved practice trenches started by its predecessors. Likewise, the 4/2nd City of London Regiment, who arrived there at the same time, found it 'a really up-to-date military station'.

Australian battalions moved to Hurdcott in 1916. Thomas Kermode of the 8th Battalion noted in his diary that in December:

> We had practice in real trenches with dinkum bombs & catapults throwing bombs, we were to have practised one of the raids you see so much about in the papers. It is really wonderful the devices & wire entanglements that can be slapped up in no time. The bombs & explosives used are terrifying in their intensity…
>
> A mock trench raid. Lieut Taylor who is in charge of bombing school let me dig a mine. I chose the man I wanted for a mate & my old mining experience

Hurdcott

stood me in good stead. A fatigue party were trying yesterday to make a hole for putting in ammunition, but had no idea ... But I knew what to do & made a success of it. Loading & firing it. Colonels, majors & all the heads about. Bombs, rockets, machine guns, etc. Men advancing in the mud with sandbags around knees & elbows. When a rocket goes up, every man lies flat & still. Just like real war.

In March 1917 Hurdcott House became the headquarters of Number 3 Command Depot of the Australian Imperial Force, whose diary acclaimed the locality as ideal for a convalescent base and thought the huts well laid out and the kitchens 'splendidly equipped' and capable of feeding 4,000 men. The AIF 'taking-over party' arrived on 12 March and by the 15th 1,700 men had reported there. Initially, Number 5 and 6 camps housed convalescing troops, but in the autumn became a 'sub hospital', with higher categories of patients transferred to Number 7 and 8 camps. Much of the training was carried out by British instructors of the Army Gymnastic Staff from Aldershot and Devonport. The depot received men who had been evacuated sick or wounded from France and were reckoned likely to become fit for active service within three months of graduated training. Particular attention was paid to dental health, with a man needing to have ten sound teeth in each jaw to be passed fit for overseas training – so that he could hold the mouth of his gas helmet properly.

The original card is tatty but is of interest because it shows Australians playing their popular pastime of Two-up. The photographer has attempted to 'improve' the outline map of Australia but has reversed the 'S'.

Hurdcott

The camp newspaper, the *Hurdcott Herald*, yields some interesting observations on military life and the method of grading men's fitness for service. One article explained:

> The whole aim of the depot is to bring the men to a requisite standard of health and fitness to undergo hard training in the least irksome warf [*sic* – misprint for 'way']. To this end the men are given as good a time as is consistent with the maintenance of military discipline.

On 1 March 1918, the *Herald* noted that 'OC [Officer Commanding] 10 Coy' had sent twenty-six men with measles to 'OC Isolation', who wrote a note back saying: 'I am full of Scabies at present and have no room for your Measles.' Another edition suggests that the military authorities should run motor services between the camp and Salisbury because taxi drivers were charging soldiers too much. In 1917 the rates from the city to Hurdcott (6 miles) were 2s and to Fovant (8 miles) 2s 6d (which seem quite reasonable compared with the London taxi-cabs' fare from autumn 1917 of 1s 2d for the first mile). A lady in July 1918 was told the taxi fare from Dinton Station to Fovant would be 4s for a journey of under 2 miles; despite the rain, she decided to walk.

Five huts were transformed into a camp hospital, two more for dressing wounds and another for examining new arrivals – some of whom were on crutches, so stone paths had to replace duckboards. One patient was William Duffell, who, after being gassed, arrived at Hurdcott on 24 November 1917. In his letters home (edited by Gilbert Mant and published under the title *Soldier Boy*), he wrote:

> The scene here was row after row of wooden huts & to one of these I was alloted [*sic*], together with some 50 other rather war worn diggers. Iron bedsteads with fibre filled mattresses lined the walls & a trestle table together with forms held the centre of the hut.
>
> A tea consisting of bread & jam followed by a rice pudding was readily despatched by the troops. Mugs & plates were washed up by the mess orderlies after which blankets were handed out three to a man. Soon all were curled up in bed as it had been a heavy day for most of us who were not yet very strong.

Duffell was at Hurdcott for eight months, during which he acted as 'offsider' (an Australian term for assistant) to the camp barber, lathering the beards of those willing to pay 3d for a shave and 6d for a haircut. He was meant to be paid in chits from the orderly room, but his customers ignored camp regulations about handling money and tipped him. He also did a stint as a cook's offsider, which provided him with much richer food than the ordinary camp patients enjoyed, but he resented being a mess orderly for the officers, whom he regarded as being 'young puppies who are no better than yourself'.

Attractions at Hurdcott included a cinema, YMCA, Red Cross facilities and twice-weekly concerts, with the depot's own concert party, 'The Kangaroos', busy with bookings at other camps, such as Codford, Sutton Veny and Sand Hill. Another popular troupe were 'The Boomerangs'.

In November 1917 Number 4 Command Depot moved from Codford to Hurdcott and in August 1918 Number 3 Command Depot was ordered to prepare to disband, dividing its men between Number 1 Command Depot at Sutton Veny and Number 2 at Weymouth, but the Armistice appears to have prompted a change of plan.

By January 1919, with no longer a need to rehabilitate men for fighting, the command depots had all but ceased to function as such and were receiving men from France on their way back home, most of that month's 3,095 arrivals being in that category.

Awaiting repatriation at Hurdcott in 1919, Number 4 Squadron of the Australian Flying Corps had a mascot who attracted much press publicity. In Germany shortly after the war's end the squadron had adopted a French lad aged about eleven, whom they nicknamed 'Digger'. He was smuggled back to England in a kitbag and taken to Hurdcott, where he was spoilt by everyone. On a trip to London his comrades paid £21 to fit him out with a uniform and extra clothing, then spent a further £12 in Salisbury on toys for him. When the squadron left for home in May, he was hidden in a hamper, which was loaded into the luggage van on the train to Southampton and then on board ship. (In 1926 Digger, now a naturalized Australian, joined the Royal Australian Air Force as Henri Heremene Tovell, having taken the name of his 'guardian' at Hurdcott, Air Mechanic Tim Tovell, but was killed in a motorcycle accident in 1928.)

When camp stores and equipment were advertised for sale in August 1919, five pianos, six billiards tables and hundreds of cricket bats were included. The camp was demolished early in the 1920s.

LARK HILL

In 1897 the Army began to buy land in the Lark Hill and West Down areas of Salisbury Plain for use as artillery ranges and a Royal Field Artillery practice camp was set up on Knighton Down, 2 miles north-west of Amesbury. In 1906 the RFA site was named Durrington Camp, distinct from Lark Hill camping-ground, which was laid out close by on both sides of the Packway track. Also established nearby were Fargo and Hamilton camping-grounds, in 1904 and 1909 respectively. A little further away two more camping-grounds were constructed near the Bustard Inn in 1903 and Rollestone Bake Farm in 1904.

Lark Hill

Sheds were erected in 1909 to house Army officers' privately owned aeroplanes, followed by hangars used by the British and Colonial Aircraft Company. This led to Lark Hill pioneering military aviation for five years.

In late 1914 Sir John Jackson Ltd started work on one of the largest camps in the country, with no fewer than thirty sets of hutments, each able to hold a battalion of 1,000 men, on the camping-grounds at and close to Lark Hill. Four more hutments were built at Rollestone Camp and were numbered 31 to 34 in the Lark Hill sequence. Buildings were also erected on the Fargo and Hamilton sites.

Rudyard Kipling visited Lark Hill in November 1914, where he was critical of 'leisurely' camp construction workers and the attitudes of other civilians:

> Three perfectly efficient young men who were sprinkling a golf-green with sifted earth ceased their duties to stare at [a fleet of Canadian lorries]. Two riding-boys, also efficient on racehorses ... cantered past on the turf. One gentleman has already complained that his 'private gallops' are being cut up by gun-wheels and 'irremediably ruined'.

On 17 December the Canadians' 2nd and 3rd Infantry Brigades moved into the new huts at Lark Hill, some of them having been involved with their completion, which had been delayed by adverse weather and a civilian workers' strike over pay. The huts were poorly built and very draughty. The men sealed the leaks with paper and cloth and went to sleep with the stoves as hot

The notorious Lark Hill mud, portrayed in the winter of 1914–15.

as possible, only to awake with the huts icy cold, leading to a deterioration in health compared with when they were in tents.

In April 1915 the War Office created Lark Hill District, an administrative area separate from Salisbury Plain District and comprising all hutments between Durrington and Shrewton, Fargo Hospital and the camping-grounds (then little used, if at all, after accommodating Canadians from October 1914 to February 1915) at Bustard, West Down and Pond Farm. There was accommodation for 35,000 men, 1,200 patients and 8,000 horses.

From 1915 many artillery units spent a fortnight or so of final training in the Lark Hill area before going overseas. F. W. Paish arrived there in November 1916 as an officer cadet with the Royal Artillery:

> our training was mainly practical. We practised our gun-drill on modern 18-pounders... we drove our vehicles for considerable distances round the largely empty countryside, and, above all, we learned to use our maps, both for finding our way and for laying guns ... The month's course at Lark Hill concluded with firing practice with live ammunition. Each cadet in turn had to act as forward observation officer, while the others manned the guns, taking different positions in turn. As gun crew, I heard for the first time the noise of a gun being fired and the astonishing shriek of the departing shell, and saw the length of the recoil.

Infantry units from other training camps – some as far away as Winchester – also spent a week or two at Lark Hill so they could practise on the rifle ranges.

The camp included a Royal Flying Corps training centre, where men were 'handed over to the Military to take the rough edge off before the Flying Corps will have anything to do with us. The Military hate us because of our pay and being a privileged corps established over their heads,' noted one air mechanic based there in 1915–16.

By mid-1916 Lark Hill was starting to lose its reputation for mud everywhere as conditions were being much improved. 'The roads running between the rows of huts are of substantial metal, kept in a state of good repair,' Frederick Crawford wrote home to his parents in New South Wales. 'Macadamised paths make the pedestrian traffic of the camp cleaner, drier, and more comfortable than it otherwise would be; drains where necessary run off the flood and soakage waters, ensuring in no small degree the health of our boys.'

From mid-1916 Australian troops were the major occupants of Lark Hill. On 22 February 1918, with the main New Zealand depot at Sling overcrowded, 130 New Zealand staff moved to Number 4 Camp to prepare for the arrival of the 33rd Reinforcement from home; 600 men were expected, 1,100 arrived. Prone to illness after a voyage through the tropics to the cold of the Plain, they were not allowed to visit other camps as a precaution against infecting other troops. When in July men of the 35th Reinforcement were afflicted by measles, part of Lark Hill was turned into 'a measles internment camp'; those on leave were limited to a 5-mile radius and had to wear a yellow stripe on each arm to denote they were in isolation.

In 1919 at Lark Hill, as elsewhere, soldiers became impatient with continuing drill, Army discipline and delays in demobilization. Most huts were used only by troops based there for a few weeks and with little regard for their upkeep they took on a semi-derelict, shanty-town appearance and were vulnerable to vandals and scavengers. But with the influence of artillery in the Great War (in which it was responsible for more than 58 per cent of British casualties), the future of Lark Hill and its ranges was in little doubt and the School of Artillery was established there, combining various gunnery schools, in 1920. The wartime huts were gradually replaced, married quarters being among the first new buildings, and photographs of the early 1920s show a far more civilized appearance.

The wartime hutments at Lark Hill Camp extended dangerously close to Britain's most important archaeological site, Stonehenge. Their southern perimeter was just 730yd from the monument itself but closer still to the Avenue, and Camps 8 and 9 were next to the north-east edge of the Cursus. (The Avenue and Cursus are ditches and banks of archaeological significance forming part of the Stonehenge site.) Sewage trenches from Lark Hill threatened the Cursus, whose western end was also damaged by ploughing. Trenches were dug on its north boundary and nearby was a bomb-throwing station, perhaps the source of the 'mines' that in April 1916 were blamed for cracking a recumbent stone at Stonehenge. This seems a little fanciful, and one would think that the upright stones, some of which were supported by wooden props, were more at risk. Rather more substantial were concerns about troops using a right of way running through the ditch and bank surrounding Stonehenge and turning it into a road. Later the authorities blocked it with barbed wire, but in June 1918 it was reported that this had been broken through. With the return of peace, the original right of way was legally closed, but only after a great deal of legal activity and Treasury quibbling over expenses charged by local professional people, such as surveyors.

Government officials and Southern Command staff appear to have been genuinely concerned about threats to the ancient landscape, the latter regularly including in orders an admonition that excavations into mounds and barrows be avoided. The problem was not helped by the constant turnover of troops and in 1918 one officer admitted he had no idea that a barrow was a burial mound. Concerns were allayed in March 1918 when Lieutenant-Colonel William Hawley, a retired Royal Engineer and Fellow of the Society of Antiquaries who lived locally, was appointed to report to the authorities damage to ancient monuments, and, hopefully, any threat to them. (From 1919 to 1920 Hawley was employed by the Office of Works to excavate the Stonehenge site.)

Lark Hill

This section of a 1in 1920 Ordnance Survey map shows the Lark Hill complex. The Military Railway runs from Amesbury through Lark Hill (with a branch to Rollestone) and past Fargo Hospital to Stonehenge and Lake Down airfields. The map wrongly transposes the names of Lark Hill and Durrington camps.

NETHERAVON

In March 1898 the War Office acquired the Netheravon Estate, six years later opening a 'school of instruction for cavalry' at Netheravon House, close to the main Marlborough–Amesbury road with the purpose of teaching officers and NCOs the best methods of training men and horses. The syllabus covered equitation, practical horse management, practical instruction, tactical exercises, reconnaissance, skill at arms, field engineering and strategy.

Huts were erected in the grounds in 1907, about when an indoor riding-school and stables were also built. At the end of 1912, the Royal Engineers' Air Battalion took over some of the quarters and lived there while constructing aircraft sheds and other buildings on the downs above the village.

Canadian troops established a veterinary hospital at the cavalry school to cope with horses suffering from the wet, muddy conditions of the winter of 1914–15. The riding-school itself was turned into a temporary hospital for the men.

During the war the first floor of Netheravon House housed a library, reading and writing rooms and, later, a mess run by members of the Women's Army Auxiliary Corps. Cadets were treated as troopers and had to take off their officers' badges and wear cadets' white hat bands. As well as instruction in horse-riding, the course included the Hotchkiss gun, musketry, bombing and entrenching. A day was spent at Tidworth going through the gas chamber there to accustom the cadets to the gas that they might encounter in France.

The cavalry school transferred to Northamptonshire in 1922, when a machine-gun school replaced it at Netheravon.

PARK HOUSE

Though Park House, or sometimes Parkhouse, Camp (map reference 220448) was in Hampshire, close to Shipton Bellinger, it was only just inside that county and halfway between Tidworth and Bulford, therefore is integral to Wiltshire's military history. It was established as a camping-ground in 1900.

The 58th Infantry Brigade was based there under canvas shortly after the start of the Great War, one of the first units to arrive being the 8th Cheshire Regiment, who detrained at Tidworth and marched to the camp, 'having drawn equipment of a sort from the station'. A modest meal of bread and tea was purchased from a grocer's cart.

Other early arrivals were the 4th South Wales Borderers, who comprised 1,000 raw recruits, three officers and fifteen NCOs from the 1st Battalion. Many tents had no floorboards and a storm blew down the YMCA marquee, library and band tents and many bell tents. When most of the battalion moved into billets at Basingstoke on 7 December, a hundred men were left behind to help build huts, a prelude to the 4th becoming the divisional Pioneers – it had a high proportion of officers and men with mining and engineering backgrounds.

The huts were ready for occupation early in 1915. A barn in Vigors Field on the edge of Shipton Bellinger was converted into Arnold's Cinema.

When the 2/5th Gloucestershire Regiment arrived there in 1916, its first impressions as recorded in the regimental history were not encouraging:

> the weather was bitterly cold; there were no palliasses, there were no fires and no light as the electricity failed on the night of the Battalion's arrival. The men were put into huts, given rations and three blankets a piece and left to get what sleep they could on bare bed boards ... despite the discomforts and the tedium of training, Parkhouse Camp had quickly become its home, and though doubtless no one was sorry for a change of venue, yet when the day of departure arrived, there was a feeling of reluctance to leave the Plain and the memories it held in its undulating folds.

On one exercise the 2/5th was judged to have won a hill that it had attacked, whereupon the men lay down, oblivious to the threat of counter-attack and making no attempt to consolidate their position. Adding to the unreality, a market woman from Tidworth strolled around selling oranges.

Park House Camp in 1915. Primitive bathing facilities can be seen in the bottom left-hand corner.

The war poet Ivor Gurney was a member of the 2/5th and in February 1916 wrote from Park House:

> This is a pretty place, far better than expected ... And huts are far better than tents. There is a stove going all day ... and make things look more cheerful, and the men are noisy and happy – they are bellowing popular songs, in a robust but sentimental fashion – a good lot of chaps.

Bernard Montgomery, the future field marshal, was a brigade major at the camp from August 1915 to January 1916. 'This was a very pleasant spot and some really good work was put in,' he recalled.

In May 1917 Clifford Allen, one of the most resolute of conscientious objectors or 'Absolutists', was sent to cells at Park House to make mailbags. Conscientious objectors prepared to do such menial tasks as this and cookhouse and cleaning fatigues were sometimes attached to Army units. Allen believed that this freed other prisoners to do work that might help the war effort, something to which he was strongly opposed. He was court-martialled on 25 May, his defence later being published by the No-Conscription Fellowship in a pamphlet, *Why I Still Resist*. Paraded before troops at Park House, he was sentenced to two years' hard labour at Winchester Prison. (Allen was Treasurer and Chairman of the Independent Labour Party between 1922 and 1926 and was created Lord Allen of Hurtwood.)

Park House became a depot for the Australian Imperial Force, housing training battalions as well as engineers and signallers, its Army Service Corps and Army Medical Corps. Some of the Australians who arrived from home with mumps were treated at the camp hospital, where they were impressed with the treatment. In early 1917 a Convalescent Training Depot was set up, providing military training for soldiers with syphilis, followed in July by a similar arrangement for those with gonorrhoea. The soldiers transferred daily from the Dermatological Hospital at Bulford. (In the summer of 1917 the Depot was temporarily at Bulford.)

In April 1918 the commanding officer of the Australian Provost Corps singled out Park House as a camp where discipline was not as good as it might be, suggesting that supervision was lax – this after his men had been hooted and abused when performing their duties on ANZAC Day, when celebrations had become boisterous. On this or another occasion, two Australian military policemen were saved from violence at the hands of their own troops only by the armed intervention of a British officer and his wife living in Manor Farm House.

Following the departure of the Australians after the war, the camp became derelict, with local children enjoying playing in the practice trenches and other earthworks constructed by the troops. Most structures were demolished and by 1923 only a few were left.

PERHAM DOWN

Perham Down, between Tidworth and Ludgershall, was part of the Tedworth Estate bought by the Army in 1897. (Its name is sometimes rendered as 'Perham Downs', especially by Australian soldiers.) It is just inside the Wiltshire boundary, but the camp was sometimes identified with Andover in Hampshire, as in the case of Post Office services and reports made by the Australian commandant in 1917. A camping-ground was established in the summer of 1909 (map reference 258491), in June hosting eight Militia regiments totalling 3,000 men. On 15 August 'Red Army' troops started to assemble there for the forthcoming manoeuvres, most having detrained at Ludgershall Station. They moved off to Salisbury after an inspection by the Duke of Connaught.

Of the early Wiltshire sites, Perham Down, together with Windmill Hill, was the most popular, being a mile from Ludgershall Station, which not only meant short marches when arriving at and departing from camp, but also facilitated off-duty visits by train to Savernake Forest, Marlborough, Swindon and Andover. The Adelaide Baths in the last town proved popular after the spartan washing facilities in camp, which were essentially a 100,000-gallon bathing pond. Ludgershall itself was an easy walk away and though small it

Perham Down Camp may have been Spartan, but before the war Blanchet of Tidworth indulged the troops there with his hairdressing and shaving services.

had several public houses and, in the war, its own YMCA facilities and the New Café, which provided 'good feeds'.

On 22 May 1913 General Sir Horace Smith-Dorrien reviewed 7,600 troops at Perham Down, with aircraft of the Royal Flying Corps taking part for the first time in an Army display, the RFC having been formed the previous month. A similar Grand Review took place the following year. Such spectacles attracted thousands of spectators, who, together with so many troops, must have placed great pressure on the limited facilities of Ludgershall.

On the outbreak of war the site was used by some of the new Kitchener battalions. It was quite common for marquees to blow down in the winter gales, but the men are said, in the spirit of the early days of war, to have borne their discomfort splendidly, though one nickname for the spot was 'Perishing Down'. Troops under canvas were there until at least 8 December, when they were moved into billets. The 10th Worcestershire Regiment expected to go to Leamington, but after an outbreak of mumps found itself destined for Weston-super-Mare – a rather more congenial place to spend the winter than in the mud of Salisbury Plain.

No doubt the 6th Wiltshire Regiment, a proportion of whose men would actually have come from the county, were hoping to be sent to near their homes, but they ended up in Basingstoke (still reasonably convenient, being 40 miles from Salisbury) before returning to Perham Down on 24 March. By then, huts for 5,000 men were nearing completion. The 7th East Lancashire Regiment, which had formed at Tidworth Pennings the previous autumn, arrived there in April. H. W. House, then a second lieutenant, recalled:

> training became more exacting. We were gradually being supplied with proper uniforms and weapons and equipment etc. Of course, privately speaking, we really lived for week-end leaves – Saturday 1.00 p.m. till Sunday midnight; but one was lucky if one got leave on two weekends out of three, and often less than that – there were always the Adjutant's instructions to be borne in mind, with odd jobs to be done, and one seemed to spend endless Saturday afternoons bicycling to Andover and waiting in a queue to get one's hair cut short enough to meet the Adjutant's approval, or buy the coloured gloves that he passed! If one went on leave one had to bycycle [*sic*] to the station, quite a distance, and bicycle back to camp off the last train. However we managed to enjoy the short week-end leave very much, though the trains to and from London or elsewhere were not very good.

When the 7th Lincolnshire Regiment left the camp to move overseas in July 1915, the men serenaded their colonel outside the officers' mess marquee, their refrain commencing 'you've got a kind face you old bars-tard – you ought to be bloody well shot'. This appears to have been taken in good part, but less forgiving a few months later was the commanding officer of the 16th

Middlesex (Public Schools) Regiment, when on their last night some of his men wrecked the canteen; he paraded the battalion and required all its members, whether involved or not, to pay for the damage.

The 10th Lincolnshire Regiment (the 'Grimsby Chums') found the camp 'very insanitary' in August 1915 and was glad to leave 'beastly unhealthy Perham Down' for Sutton Veny.

In 1916 Australia's Number 1 Command Depot was established at Perham Down, accommodating 4,000 men; it moved to Sutton Veny in October 1917. One section was ominously named 'the Hardening and Drafting Depot', possibly an early name for the Overseas Training Brigade that was at Perham Down from June to October 1917. As the names suggest, the aim was to toughen up men, mainly recent convalescents, for combat.

The camp was rather too close to the trench-mortar range, explosions of which disturbed some 'shell Shock cases sent here for convalescence'. One soldier wrote home that 'rat-hunting is a nice pastime here – big long brutes they are about one foot long – and we hunt them with our entrenching tools'.

One issue the Australians had with the British authorities was the way they charged for alleged damages. 'The huts ... are badly constructed and cheaply and quickly run up [built],' lamented the depot commandant in September 1917:

> The inside linings are matchboard and three-ply veneer, and in many cases purely cardboard ... the iron in many cases has rotted through within three years; the chimney stoves burnt through in one season, and the uprights supporting large sheds are both inadequate and constructed of soft wood.

When the Duke of Connaught visited the camp the Australians paraded only after a lot of persuasion from their officers and among rumours that they would 'count out' the Duke. When he appeared, they called out '1, 2, 3, 4, 5, 6, 7, 8, 9, 10, out you Tommy Woodbine bastard,' then immediately down-counted '10, 9, 8, 7, 6, 5, 4, 3, 2, 1, you're in. You're a bonza [Australian slang for first-rate chap]. You're one of us.'

By February 1919 the 1st South African Reserve Brigade was at Perham Down waiting to be shipped home. The other major occupants then were the Royal Army Ordnance Corps (RAOC), but their numbers were insufficient to dispel the derelict air. By February 1920 only a hundred RAOC men were there; many of the five-year-old huts were in increasingly bad condition, with one visitor being concerned that the costly stoves were rusting away.

However, unlike most camps built in Wiltshire during the war, Perham Down did have a future. The Anti-Aircraft Artillery School transferred to it in 1921. Next year the 5th Battalion of the Royal Tank Corps moved in and a post-war programme of reconstruction began, with many of the wooden huts being replaced by brick buildings.

POND FARM

Pond Farm, near Market Lavington, was one of various farms taken over by the War Office as part of its acquisition of the Plain. A camping-ground was laid out 1,000yd to the north-east (map reference 055535) and from about 1905 was used by mainly Yeomanry and other mounted units during the summers. It was the most isolated of all camps on Salisbury Plain, accessible from the north only by roads up a steep escarpment forming the Plain's boundary. The roads to the south were flatter, but that route was 11 miles from the station at Amesbury compared with the 5 miles from that at Patney & Chirton. Pond Farm itself appears to have been demolished soon after the Army acquired it; some such structures on this part of the Plain were hit by artillery shells, though whether they were deliberately targeted or accidentally damaged is not clear.

The camp was used by several exotic units, such as the 'London Roughriders' (the City of London Yeomanry) and the King's Colonials or 'Worldwide Empire' regiment, which had been raised in 1901 from colonials living in the London area. Their original outlandish uniforms were toned down by a new commanding officer in 1904, but even in 1913 the officers wore slouch hats with large drooping plumes of cocks' feathers.

In October 1914 elements of the First Canadian Contingent detrained at Patney & Chirton Station over several nights and marched, usually in the dark, the 5 miles to Pond Farm Camp. Of the various sites at which Canadians were based during the very wet winter that followed, Pond Farm, remote and high on the Plain, was the bleakest, though with water draining off the high ground on which it was sited it was not quite so muddy as others. The tents were without boards for two weeks and the men had to sleep on the wet ground. There was such an epidemic of coughs caused by the weather that the condition became known as the 'Pond Farm Particular'. Even unluckier was Armourer-Corporal William Ogden, who was accidentally shot when repairing a jammed rifle.

On 12 December *The Times* reported on the units camped there:

> Round the entrance to the headquarters tent in both regiments is a veritable lagoon of slime, across which no amount of road-making with plank and wattle hurdles and bundles of cut furze-bushes will enable one to reach the tent without getting into mud up to one's bootlaces. Nonetheless, on the whole, this part of Pond Farm is dry, and it is sometimes immensely invigorating in the thrust and bluster of the wind (though it is rather hard in tents), and the health both of men and horses has been excellent.

Though the correspondent did allow that a recreation tent was needed at night, his report was far too sanguine, especially about how the horses were coping with the conditions. In reality, they were not, needing proper shelter

George V showed interest in these primitive armoured cars at Pond Farm Camp in November 1914, but Lord Kitchener disagreed when the King suggested they would prove useful. They were designed to carry troops and provide a mobile reserve of firepower.

even more than the men, but having to make do with canvas slung across crude frameworks and the branches of trees.

At issue in December 1914 was the state of the road from Patney & Chirton Station to Pond Farm up Red Hone Hill (now better known as Red Horn Hill), up which 80 tons of supplies were hauled daily. Though the military authorities admitted liability for the road from the top of the hill to the camp, Devizes Rural District Council felt that since the Army had created the problem, then it, not local ratepayers, should pay for mending all of it. Eventually the council agreed to spend £2,470 on repairs.

After the Canadians' departure in early 1915, Pond Farm Camp appears not to have been used again. No doubt the reasons included its isolation from the other Salisbury Plain camps, the difficulty of supplying it with stores and an understandable preference for hutted quarters. It was omitted from a 1921 Southern Command list of camping-grounds and eventually became part of the artillery ranges' impact area.

PORTON

Porton Down, close to the London & South Western Railway and Porton Station, proved a convenient assembly point for troops, with 50,000 soldiers massing there in September 1898. In 1899 the Royal Engineers set up a base near Porton village while they installed telegraph lines between three new camping-grounds (Bulford, West Down and Perham Down) and Tedworth House, their Salisbury Plain headquarters. Until the construction of the Amesbury & Military Camp Railway in 1902, Porton Station saw several mass arrivals and departures by soldiers taking part in manoeuvres.

On 22 April 1915 the German Army used chlorine against the Algerian Division of the French Army and, two days later, against the 2nd Canadian Brigade (which until February had been training on Salisbury Plain). On 1 May men of the 1st Dorset Regiment were the first British troops to suffer a gas attack. The War Office immediately investigated the best ways both to counter the new weapon and to exploit it.

As a result, early in 1916 an initial 2,886 acres near Porton were acquired by the Government and converted into the 'War Department Experimental Ground, Porton'. Pedantically, it was closer to Idmiston (and Idmiston Down) than Porton, after which the site became known. As with other camps, the nearest railway station rather than the nearest village may have been a determining factor in choosing the name.

A hundred cylinders of hydrogen sulphide were delivered in March 1916 and two Army huts erected. Later, eight rows of eight huts, each capable of housing fifteen men, were put up, with 400 men of the 14th Labour Battalion, Devonshire Regiment, forming the initial construction force, which later doubled in size. The 6th of April saw the arrival of six civilians with experience in the use of self-contained breathing sets in mine rescues; they were paid 2s 6d a day for their scientific experimental work of 'sampling the [gas] cloud' as it was discharged. They were followed five days later by thirteen more civilians who were to be trained in opening hydrogen sulphide cylinders.

In mid-1916 local councillors were expressing opposition to a wire fence around the establishment and later were concerned about local rights of way being stopped up. It would seem that they did not appreciate the nature of the work being carried out. Civilian workers were committed to silence under the Official Secrets Act, though there must have been much gossip and speculation in the local villages about what was going on.

Later in 1916 the site's name was changed to the 'Royal Engineers Experimental Station', reflecting the predominance of Royal Engineers on the staff, which also included members of the Royal Army Medical Corps and a Royal Artillery experimental battery. A meteorological station was established to help the scheduling of gas tests, which depended on suitable weather; members of the Chemical Advisory Committee who visited Porton to view the experiments often had to wait several days for the right conditions. To transport workmen, fuel and stores, a light railway was built from Porton Station.

Circular gas trenches were constructed on Idmiston Down by Chivers of Devizes in 1916. One, 400ft in diameter, enclosed another that was 200ft

The narrow-gauge railway being constructed at Porton late in the war.

across. The unsupported sides collapsed that winter and the trenches were abandoned after very little use when it was realized that the effects of gas needed to be measured over miles – a greater distance than allowed for by the trenches. An Anti-Gas Department was set up in 1917 to research ways of protecting troops in trenches and dugouts, with the evaluation of respirators being a priority. An artillery range was established in 1917, with guns firing from several locations, including immediately beneath the early Iron Age earthworks of Figsbury Rings. Several structures were erected there, with four concrete blocks with curved sections at the rear enabling guns to be positioned accurately.

That same year, the Trench Mortar Experimental Station was established at Winterbourne Gunner, or 'Porton South'. Responsibilities included research into pyrotechnics, mortar and grenade signals, Zeppelin darts, body armour and barbed-wire entrenchment, as well as trench mortars. Wartime experimental work at Porton South had no connection with chemical warfare and the camp was virtually a separate entity, though its personnel came under the Porton North commandant for administration and discipline.

Crop and sugar-beet production provided fodder for the variety of animals that were the subjects of experiments, goats being particularly favoured because their respiratory systems resembled a human's. At first, the animals were kept at Porton Down Farm, but the noise of guns disturbed them and they were moved to Arundel Farm, west of Newton Tony, on a site separate to the main establishment. It is said that monkeys were released during the Armistice celebrations, frightening local people for some time until they died off.

By the end of 1918 the station covered 6,196 acres. A headquarters block was under construction on the main site and the total staff was reported to be some 1,500 officers, other ranks and civilians. The number varies with different accounts, but civilian workers appear to have comprised between a third and a half of the complement. The scientific staff had military status, though their modest ranks seemed hardly to match their academic qualifications. Women of Queen Mary's Army Auxiliary Corps, providing mainly typing and shorthand services, were quartered at the Grange, Idmiston.

The station conducted its work in near-isolation. It was not in direct touch with other countries, nor with the British Army in the field, and there was no liaison with artillery units in France. Visits of irregular frequency were made by French, Italian and United States officers to witness individual experiments. One American lieutenant was stationed as a liaison officer at Porton for a short period and two others conducted physiological research there in 1918.

The camp railway had not been completed by the Armistice and in December 1919 there were allegations in the local press of wasted effort put into it and the camp's construction. It was alleged that the workshops had been used mainly to repair officers' cars. The criticism was probably prompted by the station being all but deserted and by reports that it was costing £2,500 a year to maintain while its future was decided. But when the League of Nations endorsed the legitimacy of chemical warfare in 1920, Porton Down's role was assured and a rebuilding programme started. In 1921 the Government paid £7,140 for 913 acres of the Manor Farm estate at Idmiston, much of which it had already rented and which was littered with various types of trenches.

Land at Porton South was used by the Ministry of Munitions' Trench Warfare Research Department and it was there that the curious Pedrail Landship was tested. Looking in its incomplete state like a single-deck tramcar, it was 40ft long and was intended to be an armour-plated mobile flamethrower equipped with three heavy machine guns. Initially it ran on the only English form of caterpillar track, known as 'Pedrails' – cumbersome blocks attached to a belt.

By mid-February 1916 the base chassis had been made by Stothert & Pitt of Bath, but there were doubts that the completed vehicle would be proof against machine-gun fire at close range, as well as concern about its weight of 35 tons. It was decided to take it by road to Porton for extensive trials, but only after hydraulic steering had replaced a manual system. By 22 July it was ready to leave. However, the 'Pedrail feet-mechanisms' proved unsuitable for roads and, it is said, at Bratton had to be replaced with those of stronger steel and different design by a local agricultural engineering company, R. & J. Reeves & Son. Though this company was long-established and very experienced – it had been represented at the Great Exhibition of 1851 – one wonders if it would have had the expertise to do such work. But in 1999 Jean Morrison of Bratton recalled a Reeves worker, George Francis, telling how the company was asked to check the Pedrail's tracking mechanism and see how it could be improved, employees being sworn to secrecy and working behind locked gates.

The vehicle was tested on the steep track, then unsurfaced, up to Bratton Castle. Reeves had supplied wheeled water tanks for the Army in the drought of 1911 and it agreed with the War Department that in any correspondence or in answer to any questions the Pedrail should be referred to as a 'tank'. (In November 1915 the words 'water carrier' had been suggested as a 'disguise' for other forms of tracked vehicles then being developed, with 'tank' being agreed by the two assistant secretaries of the Landships Committee on Christmas Eve, 1915.)

When the Landship finally reached Porton, it acquitted itself 'quite satisfactorily', but there were doubts about its ability to cross trenches and it never saw active service. An initial order for twelve at £3,000 each was cancelled.

Perhaps the first appearance of a conventional tank in Wiltshire was on 5 March 1917, when one, accompanied by a captain and four men, was displayed in Salisbury's Market Place to promote the sale of war bonds.

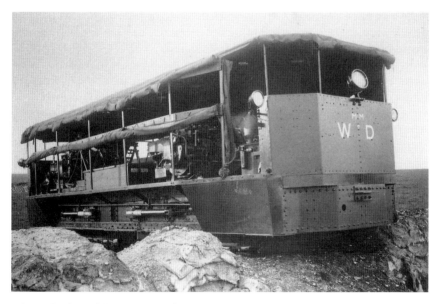

The Pedrail Landship being tested at Porton. Though its cumbersome 'pedrails' – blocks attached to a belt – cannot be seen, the low clearance below the front of the machine appears insufficient for crossing rough country.

ROLLESTONE

A tented camping-ground was established close to Rollestone Bake Farm and east of the Salisbury–Bustard road in 1904 (map reference 098448). With the outbreak of the Great War, four hutted camps were erected, with 500 men from Sir John Jackson Ltd starting work in mid-October 1914. The camp's closeness to ranges meant that it was much used by artillery units, some there for only a couple of weeks of final firing practice before going overseas.

Norman Tennant of the Royal Field Artillery noted in *A Saturday Night Soldier's War* that the area 'was a vast sea of mud and wooden hutments and it was here that I engendered a lasting hate for the inimical menace of army camps'. Others were equally unimpressed with Rollestone.

The camp was taken over in 1916 by Australian training battalions, some of whom lived in tents during the summer. In 1917 an Australian from the camp was the first locally to be accused of assaulting the police; he received two sentences of six and three months' hard labour, to run concurrently. Later that year another Australian was killed when being trained to use a Lewis Gun; one shell in its magazine of dummy ammunition proved to be live. Ivor Williams of the 21st Battalion thought Rollestone 'an awfully out of the way place', but enjoyed several visits to Rollestone Manor on the edge of Shrewton, where he was hospitably received, the house and grounds apparently being made available to local soldiers.

On the western side of the Bustard Road the Royal Flying Corps established Number 1 Balloon School, which continued after the war as the School of Balloon Training and was mainly concerned with training observers. The original huts were demolished shortly after the end of the war, though balloon hangars erected in 1932 still stand prominently above more recent military structures.

SAND HILL (LONGBRIDGE DEVERILL)

Late in 1914, Canon John Brocklebank, writing in the parish magazine, warned his fellow-residents of Longbridge Deverill, 2 miles south of Warminster:

> A great change is going to take place in our life; our quiet village is to be inundated and we are to have a camp of 4,000 soldiers settled down on Sandhill. Sleepy hollow we shall be no longer. The changes come as a great surprise to

Sand Hill (Longbridge Deverill)

us, as great as the outbreak of war seems to have been to a sleeping and careless and neglectful Government and a peace persuaded nation. We are going to learn the lesson of our lives.

The *Warminster Journal*, having regularly featured the tented camps established in the Wylye Valley within weeks of the war's beginning, barely mentions the hutted one at Sand Hill (occasionally spelt as one word). By the time its construction had started press censorship was restricting references to military works and activities.

The camp was duly built (map reference 874413) and was variously called Sand Hill (which was not meaningful to anyone unfamiliar with the locality) and Longbridge Deverill (which was more informative, if a mouthful). It accommodated only infantry units. Administratively its constituent camps were numbered 11 to 14 and were linked to numbers 1 to 10 at Sutton Veny, 3 miles away; in 1915 John Bates gave his address as 'Sand Hill, Sutton Veny', an example of the identity of a satellite hutment being subsumed under that of the major camp in the neighbourhood.

In the summer of 1915 the Oxfordshire & Buckinghamshire Light Infantry judged Sand Hill to be 'by no means such a comfortable camp as Fovant' (where the unit had been based previously) and made a puzzling reference to the huts there being 'older', when in fact they had been erected only a few months previously, at the same time as those at Fovant:

> The huts were smaller, older, closer together, and in many cases none too impervious to rain. There was a great lack of suitable parade grounds. The country round, however, gave promise of less exhausting field days, for it was not nearly so hilly. There was, too, another great attraction at Sandhill which was lacking at Fovant – namely the proximity to the country town of Warminster.

During their time at Sand Hill in the last quarter of 1915, the four 'Tyneside Scottish' battalions of the Northumberland Fusiliers produced the *Sandhill Lyre*, whose contents often reflected the frustrations of waiting to be sent to the Front, one piece of verse being entitled 'The Forgotten Brigade'. The Fusiliers were part of the 34th Division and the perceived delay was attributed to its artillery not being ready for active service. In fact, the Fusiliers spent only a couple of weeks longer than was usual at a training camp, possibly as a result of its intended destination being belatedly changed from Egypt to France.

As they approached the camp in January 1916, members of the 2/15th London Regiment (the Civil Service Rifles) appreciated its comfortable appearance, until they saw the mud outside the huts, then inside them the filth left by the previous occupants – a Tyneside Scottish battalion. The Rifles quickly participated with other units of the 60th (London) Division

Sand Hill (Longbridge Deverill)

in a competition to see which had the smartest quarters. Luckier were men of the 2/16th London Regiment (the Queen's Westminster Rifles), who arrived at the same time as the Civil Service Rifles; Bert Pedler reported that 'the huts are clean and fine and large. They are raised about 2 feet from the ground and each one contains a … slow combustion tortoise stove.' At first, the new arrivals had to sleep on the floor, but later they were supplied with straw mattresses, pillows and four blankets each. Beds were issued later.

Rivalry between units was common. It was encouraged in sporting events and tolerated in spontaneous encounters provided matters did not get out hand, as sometimes happened off-duty. A grand snow fight involved the London Regiment in the winter of 1916 when its 2/15th and 2/16th battalions at Sand Hill were attacked by the 2/14th, based at Sutton Veny. At first, the 2/14th – the London Scottish – led by its subalterns and encouraged by bagpipes, enjoyed the element of surprise but were forced to retreat, some in small groups. These were surrounded and made to bend over with their kilts lifted over their heads, snow then being applied to their hindquarters. The main body of Scots retreated towards Sutton Veny, only to be

The interior of a hut at Sand Hill Camp in 1916. The sender of this card writes that 'you see the boards beds & blankets that is how they have to be folded every morning they are turned out into the room at night.'

attacked in the rear by the 2/18th, the London Irish. That night a collection of 'war trophies' – bagpipes, glengarries (brimless hats worn by some Scottish units) and sporrans – were displayed in the YMCA hut at Sand Hill before being returned the next day.

At Sand Hill in 1917 the 2/9th London Regiment (Queen Victoria's Rifles) brought together what it termed its 'odds and sods' into a headquarters company, its officers wondering if they were the first to do this; certainly they considered it a novel concept, though at first it was not popular with company commanders whose individual strengths were reduced. The new company of a hundred men comprised the machine-gun section, transport personnel, medical and quartermaster's staff, signallers and pioneers.

Four miles from the camp, Cold Kitchen Hill, west of Monkton Deverill, was an abiding memory for many men, some later saying a night march over it was worse than a bad patch at the Front.

As at most camps, bathing facilities were minimal, with nearby Crockerton Pond proving popular in the summer. A camp labourer drowned there in June 1915.

Australian troops arrived at Sand Hill in October 1917, when their Overseas Training Brigade was established there to train soldiers who had recovered from illness and wounds for a return to active service. The following January, Tom Gardner, after recovering from rheumatic fever, was there to be kitted out and was then ready to move away at any time. 'We get physical jerks, bayonet fighting, bomb throwing and all the other things that are essential for the killing of a fellow man, but the work is not hard here like the training camps because here they are all seasoned troops who have been through the actual thing,' he wrote to his sister. (Recurring ill-health meant that he was to spend the rest of the year in hospital before going to France. Further illness led to him being returned to Australia and discharged from the Army, only to drown when bathing thirteen days after the Armistice had been declared.)

When the camp wound down with the return of peace, Canon Brocklebank reflected on the silence in the village:

> How we miss the noise and bustle; no reveille at 5.30 a.m., no trains of mules, no interminable practising of the same old tune upon the bugle, no swinging tramp of men through the village, no military march music, no dust, no rattle of rifle fire ... The soldiers had found a way with our affections so surprising, so unexpected, the pleasure was all the more.

The Sand Hill huts were among the first in Wiltshire to be demolished after the war.

SHERRINGTON

A camping-ground at Sherrington was one of those hastily laid out in the Wylye Valley after the outbreak of war, with 3,000 men arriving on 22 September 1914 as part of the new 26th Division. They commenced their training in a motley of civilian dress and discarded uniforms of other regiments, with the divisional headquarters being the parish room built of galvanized iron.

At first, the site looked attractive, but when winter rains set in it was discovered that the spot, close to the River Wylye, had not been chosen with much foresight – though local farmers had warned about its choice. The 7th Oxfordshire & Buckinghamshire Light Infantry reported only two dry places – the officers' lines and the cookhouse:

> The camp was situated, like many others in the district, in a gently-sloping field close to the river. Had advantage been taken of the higher ground, the trials which had to be endured later from the mud might have been very considerably minimized. The local farmers, by way of encouraging us, were eager in their warnings of what we might expect when the rains did set in, and the Wylye begin to rise. But even their worst pictures were a poor portrayal of

Sherrington Camp existed only for only three months, not long enough for amenities to be introduced and the soldiers have hung their washing on a line (at the bottom of the photograph).

what was yet in store for us. Mud – unrelenting mud – mud everywhere. First of all, sticky and clammy, then softer and deeper, and finally nothing but a sea of a heavy, over-boiled pea-soup consistency. In vain, efforts were made to make paths. Unfortunately there was no 'stone' handy, no chalk flints in any appreciable quantity – only sundry large lumps of chalk. These were dropped into the mud in line but very soon disappeared from view, and the last state was little better than the first.

Drill was possible only until the end of October, at which point the rains forced training to be done on the high ground to the south of the camp. By November tolerance with the conditions was running out and everyone in the 'Oxon & Bucks' was relieved when they moved that month to billets in Oxford. But this was not before the canteen marquee blew down on 10 November, with many of its goods 'disappearing' in the confusion.

There were few off-duty amenities. The pubs could not cope and at the Carrier's Arms at nearby Stockton the landlord pumped beer into a bath and allowed soldiers to dip cups into it. Despite a picket of twenty men patrolling the village streets, a series of incidents led to the pub closing down, ostensibly so that extra toilets could be added – certainly the existing facilities were insufficient.

On 3 October the *Warminster Journal* noted that commercial enterprise in providing for the soldiers' wants had not yet shown itself at Sherrington in such a marked degree as at Codford, where a rash of shack-shops had sprung up. Some villagers did sell goods to the soldiers, but, with no hutted camp planned, trades people must have seen little point in establishing premises there. After the troops had departed into billets or neighbouring camps late in 1914, the village returned to comparative peace, though soldiers from Boyton and Codford were frequent visitors, either off-duty or on route marches.

SLING

A camping-ground was established at Sling Plantation in 1903 as an annexe to Bulford Barracks. At the outbreak of war, the South West Infantry Brigade of Territorials was based there and quickly moved to its designated home defence stations guarding vulnerable installations, though several battalions soon found themselves back on the Plain. A few weeks later the building of a hutted camp was started by New Zealanders living in Britain who had enlisted in their country's Army. They were soon to leave to join their comrades in Egypt, so, to continue construction, Canadian troops, including bricklayers and carpenters, worked alongside civilian contractors. In November some

Bath time for soldiers at their summer camp at Sling Plantation in about 1913. A sign orders 'NO SOAP TO BE USED', a precaution against lather remaining on the water, which was changed infrequently.

huts were handed over to the First Canadian Contingent, and 'Plantation' was dropped from the title of the camp. On the completion of building, the *Timber Trade Journal* boasted that if all the huts were placed end to end they would stretch for 6 miles.

When Francis Brett Young reported to Sling in 1915, he found it a 'a place wholly forsaken by God: four miles of mud from Tidworth and two more miles of mud from Bulford'. However, weekend walks with his wife helped him to appreciate Salisbury Plain's more attractive characteristics. When the 5th South Wales Borderers were there in March and April they were kept busy constructing roads and paths and generally making the camp habitable. Only when they moved to Perham Down could they concentrate on training.

Sling was taken over by New Zealand forces in 1916, its layout then comprising four main sections or 'lines': Auckland, Wellington, Otago and Canterbury, each with its own headquarters, commanding officer and training staff. Officially called the 4th New Zealand Infantry Brigade Reserve Camp, it comprised the First, Second and Fourth Brigade Depots (the Third being at Brocton in Staffordshire) and trained reinforcements from home before they left for France. It also retrained casualties who were regaining fitness.

'Sling unloved, bleak, and lonely,' lamented Lieutenant H. T. B. Drew in *The War Effort of New Zealand*. 'Who will not remember the bull-ring of Sling, with its bare slippery surfaces, its bleak winds sweeping across its snow-covered rifle ranges or its dreary heat, and its monotonous tasks?' Nevertheless:

> the huts were comfortable and warmed in winter, the food was wholesome, well cooked, well served on hot plates, the canteens well appointed and with good libraries and billiard tables; and the spacious Y.M.C.A. with its devoted, kindly English ladies and its cinema catered well for the leisure hours.

The YMCA hut was officially opened in March 1916 and was one of the best equipped on the Plain. There were two large rooms, one for refreshment, the other for meetings and writing, with smaller rooms for the attendants and private devotion, a Post Office counter and general store.

Of all the Wiltshire camps, Sling had the toughest reputation when it came to training, which was all the more arduous after a sea voyage of two months. It had its own 'Bull Ring', run by foul-mouthed instructors insisting on complete discipline, and was comparable to the notorious training ground of the same name at Étaples in France. The training lasted from 6.30am until 9pm, often seven days a week, and extended over three or four weeks, which eventually was judged too short to equip the men for France. Towards the end of 1917 the course was lengthened to eight weeks.

If earlier in the war waste of food had been an issue at many camps, nothing was lost at Sling in May 1917. Left-over bones, fat and vegetables were put into separate bins; for a soldier to waste a crust earned him a reprimand. The monthly reports of Australian depots had a section detailing the amount of bones sold, dripping saved and soap made from 'grease trap fat'.

On 31 October 1918 there were 4,300 men at Sling. With the war's end only days away, many of these were fortunate in not having to return to fighting, but others were not. For already there was grim evidence of the Spanish influenza that was to kill 200,000 people in Britain in the coming winter. A New Zealander wrote home on 26 October: 'The influenza is all over England and hundreds are dying of it. Several of our Reinf[orcement] have died of it and over a hundred is in the hospital. I will tell you it is a serious thing.'

After the return of peace, Sling experienced the unrest that was evident in other camps over delays in demobilizing troops. To occupy them, New Zealand soldiers were sent on occupational courses and were put to work carving the shape of a kiwi in a nearby hillside.

By 1921 parts of the barracks had been renamed to reflect recent campaigns, three lines of huts now being known as Gaza, Baghdad and Gallipoli. Slightly incongruously, the north-west part of the camp was called Sling

Cottages; they housed civilians (perhaps on a temporary basis to ease the national housing shortages, or else camp workers). Much of the original camp was demolished during the mid-1920s, with new buildings being erected as part of the development of Bulford Barracks.

SUTTON MANDEVILLE

A small group of huts was built at Manor Farm, south of Sutton Mandeville, as an annexe to the far larger camp of Fovant, a mile away. Little has been recorded of its history, though in June 1916 a conscientious objector, Thomas Ellison, was court-martialled there for refusing to put on military clothing after being called up to the 7th Devonshire Regiment. After 112 days' imprisonment at Winchester and Wakefield work camp he was ordered to report back to the 7th, was arrested at Crewe, returned to Sutton Mandeville and then to Dartmouth, his battalion having moved to Devon. He was again court-martialled and imprisoned.

Like many Wiltshire camps, Sutton Mandeville was taken over by the Australian Imperial Force. Charles Gray of the 35th Battalion arrived at Dinton Station at 1am on 30 January 1917 and marched the 'six miles' (more like 3 miles) to the camp; he and his comrades very much enjoyed the sharp air and frosty ground crunching under their feet after being cooped up on board ship. At the camp they slept till noon, when they breakfasted on bully beef and biscuits, their cooks struggling with the stoves and frozen taps until they managed to prepare some soup at 6pm.

Even in the snow, the surrounding countryside – rather more pleasant than Salisbury Plain – proved attractive to the Australians, who visited Tisbury ('a very old fashioned village built several centuries ago') and the ruins of Wardour Castle. But a meal of two eggs, a piece of bread and a cup of tea cost them 1s 6d each.

Near to Sutton Mandeville were carved two hillside badges, those of the 7th London Regiment (nicknamed the 'Shiny Seventh') and the 13th Royal Warwickshire Regiment, a works battalion. The 7th was at Sutton Mandeville Camp from January to May 1916 and had cut its badge by 6 April that year. The 13th, which evolved into the 33rd Training Reserve Battalion, was there in March 1917. The original design of the Warwickshire's antelope badge included at its head '33RD T. R. BATTN' and, beneath it '13TH T B'. These characters disappeared from later forms of the badge, the maintenance of which, together with the 7th's, ceased in the early twenty-first century because of high costs.

SUTTON VENY

Sutton Veny's proximity to the Warminster–Salisbury railway line made it a suitable choice for an Army camp when the Great War started and by November 1914 the erection of ten hutments had begun, with rail track being laid from Heytesbury Station and a water supply of half-a-million gallons of water a week arranged. As with the other camps in the Warminster area, construction was carried out by Sir John Jackson Ltd, the work giving a short lease of new life to Westbury Ironworks, which had closed a few years before. Jackson's removed the slag and transported it to Sutton Veny and other local camps, where it was used for the roads and camp railways.

Though the camp had not been completed, the first of 10,000 troops arrived in April 1915, some having to live under canvas, or, after a period of heavy rain, in a seven-storey wool store where a hazard was sheep ticks. A villager commented: 'You'd see them marching through Tytherington from Heytesbury Station, they didn't have their uniforms yet, and you could tell what they'd been – the fishermen in their jerseys, the clerks in their stiff collars.'

Some of the first huts to be built, near Cooper's Bottom, were used to house civilian internees who worked in nearby fields. Later this accommodation was turned over to prisoners of war; when the first batch arrived, one spat at the watching village women, who surged forward in rage before being fended off by guards. The huts were 'long, low wooden buildings with raised board floors, each housing half a platoon of about twenty-five men, all of whom slept on the palliasses on the floor'. The 10th Lincolnshire Regiment found 'such jolly good huts and such a charming country too'.

Fields near Norton Waters were used for exercises in trench warfare. The troops would live in the trenches for up to a week, all food being brought up in dixies, as it would be in France. It was here that the 7th Wiltshire Regiment found out how to deter the attacking force, in this case Scottish troops. Christopher Hughes recalled:

> that an empty cartridge case on the muzzle fired off with a blank was rather good fun. No man's land was soon cleared of the Enemy patrols and they sent an officer with a white flag to request us to stop it; we entertained the officer, saw him safely back; we then refreshed at their mess and they saw us safely back, and so on through the long night.

Inspections by visiting dignitaries were not appreciated. Hughes wrote that shortly after the 7th Wiltshire arrived at Sutton Veny the Brigade it was part of was:

> paraded for inspection by a General from the War Office. It poured in torrents, we paraded in our overcoats and stood in a sodden field for just short of

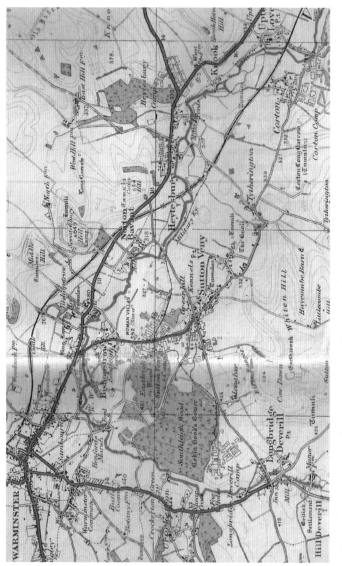

Part of a 1in 1920 Ordnance Survey map featuring the camps near Warminster and showing the track of the military railway beyond Sutton Veny to Sand Hill Camp.

two hours, then he arrived, looked over the gate at us and ordered a platoon from each Battalion to march past him on the road.

During the war I was on a number of inspections and can only remember one inspecting officer who arrived on time, Lord Allenby.

In the winter of 1915–16 elements of the 34th Division carried out a very muddy 'attack' on the trenches in front of officers from Japan, one of the nations allied against Germany. It was probably these officers who were said to have inspected the division before its departure to France in January 1916. George V was to have been there, but had been injured falling off his horse during a visit to France. The Japanese may well have been present, but it was the divisional commander, Major-General Edward Ingouville-Williams, who actually took the inspection and 'who was most awfully pleased with everything'.*

Detailed in December for Egypt and then East Africa, the 34th had been equipped with sun helmets. The orders were cancelled and the helmets returned to store, only to be reissued a few days later; the orders were cancelled again when the division was told it was going to France. (There are several other similar stories of units in Wiltshire being issued with tropical kit, only to have their orders changed and being sent to France.)

At first, staff headquarters were at Greenhill House, where the Australian YMCA later had one of the most luxurious welfare centres in the country, certainly in Wiltshire. As well as the usual reading and writing rooms, there were billiards rooms, gymnasium and squash, tennis and basketball courts and a croquet lawn. There were two 'prize rose gardens' and woodland walks. A member of the International Hospitality League arranged tastes of English home life for those going on leave and a 'snap-shots for home' department enabled men to send home photographs of themselves.

In mid-1916 there was also a YMCA refreshment tent on the firing ranges. By the end of July 1915 a church had been built and a Nonconformist chapel was under construction. Among the various welfare amenities was a United Army Board Hut. The Board catered for Baptists, Congregationalists and Primitive and United Methodists, but men of any religion or none were welcome at the hut.

When Australian Number 1 Command Depot moved to Sutton Veny in October 1917, it found the huts to be in good repair (its commandant having been critical of the state of the three-year-old huts at the previous base, Perham Down), but they 'had been left in a disgracefully unclean and insanitary

---

*Major-General Ingouville-Williams was known as 'Inky Bill' and had a reputation for being dynamic, having led reconnaissance patrols in No Man's Land. Soon after the 34th Division arrived on Salisbury Plain his car crashed into a traction engine; he fractured his skull and was unconscious for several days. He was hit by shellfire in France in 1916, becoming one of seventy-eight British and Empire generals who died as a result of active service.

This page is from a souvenir brochure of Australian YMCA facilities at Greenhill House, Sutton Veny. The signatures of staff and soldiers include that of Harold Boas, a YMCA official appointed to look after the interests of Jewish members of the Australian Imperial Force.

condition' by the outgoing British troops. There were some fears from local people regarding the newcomers' behaviour, perhaps offset by the anticipation of being able to raise prices for the better-paid Australians, such as charging from 18s to 25s to hire for a week a poorly maintained bicycle worth £3. Earlier that year Harry Patch, destined to become the last surviving soldier to have fought in the trenches of the Great War, had paid 2s 6d to hire a machine for the weekend while at Sutton Veny.

Shack-shops were built at the Quarry allotments, close to the Crockerton Road, and there were several tea-huts in a field known as 'Garden City' on the Bishopstrow Road. Early in the war village women took in soldiers' laundry, charging 1d for washing a pair of socks, 2d for a shirt and 3d for white shirts, which included repairs and replacing missing buttons. Later on, laundries were established within Sutton Veny and other camps, staffed by soldiers unfit for active service and civilian women, the latter being paid 25s a week.

A hospital opened in 1916, at one time having beds for eleven officers and 1,261 soldiers. On 10 January 1919 staff of the 1st Australian General Hospital transferred there from France to Sutton Veny. At the end of the month there were 1,003 patients, including 331 German prisoners, most of the latter being transferred to Fargo Military Hospital in late February. Most patients were suffering from Spanish influenza and many died from it. There were rumours that the cause was something even worse than this particularly virulent form of flu, perhaps even cholera; sanitation in the nearby villages was inadequate and at one time water supplies were polluted by sewage. Many of the 143 Australians who were buried in the local churchyard were flu victims, as was Jean Walker, who died in October 1918, shortly after becoming matron of the hospital. Reduced hospital facilities continued to be available at the camp until the beginning of 1920.

One problem in 1919 was the amount of refuse, notably empty tins. Eighty years before separating domestic rubbish became the norm in Britain, the hospital kitchens were dividing ashes, tins, paper (for burning) and broken glass and crockery for easier disposal in Warminster. Even old tea leaves were kept to help allay the dust in the corridors and the hospital diary described as 'most reprehensible' their being discarded into refuse bins.

The Australians established an agricultural training depot at Sutton Veny in March 1919 to prepare troops for civilian life, with courses in orcharding, wool-classing, beekeeping and sheep- and cattle-breeding; it had the additional aim of keeping them out of trouble until they could be repatriated. By September, 2,000 students had attended the depot.

At the end of July 1919 other Australian depots in England were closing, with Sutton Veny becoming the final assembling base for Australians from France (where they had been mostly engaged in reburial work). Once all the troops had departed, most of the huts were removed and the camp railway lifted.

TIDWORTH

In September 1897 the War Department purchased the 6,618-acre Tedworth Estate ('Tedworth' being an alternative spelling to 'Tidworth'), including a large mansion, for £95,000 from Sir John William Kelk, who had been trying to sell it for four years. The Army had been interested only in the land and tried to resell Tedworth House but without success, so in December 1898 it was allocated to Colonel J. M. Barklie, commanding the Royal Engineers on Salisbury Plain, and his staff, who were responsible for overseeing the building of a large barracks nearby. Later the house was the residence of successive generals commanding local troops.

Between 1899 and 1902 there were several changes of mind about what type of unit – cavalry, artillery or infantry – should occupy the barracks.

When construction started in 1902 the aim appears to have been to favour the infantry. The contractor, Henry Lovatt Ltd, first laid down railway track from Ludgershall, which continued through a civilian station on to the site of the barracks. Lovatt's employed 2,800 men at any one time on the project. To house them and their families, a temporary village was erected in 1901–02 at Brimstone Bottom (map reference 249500), close to the new railway. The buildings were in streets 400yd long and had iron walls, felt roofing and matchboard lining.

Known as 'Tin Town' or 'Navvy Village', the community had its own train service, the 'Tin Town Mail', to transport the workers to the barracks site and back, and its own stores, canteen and church, as well as an emergency hospital, paid for by a 2d deduction from each man's weekly wage. There was also a four-bed isolation hospital, which treated at least one smallpox case. (A larger military isolation hospital treating soldiers with infectious diseases such as mumps was later opened, perhaps using the same building.) Nevertheless, there was some concern about the strain that the village's inhabitants were placing on Pewsey Rural District Council's health services, though it was pointed out that Lovatt's paid a very large share of the local rates. Other amenities included a library and clubs for cycling, football, cricket and quoits.

From January 1902 to July 1904 no fewer than 26,000 men turned up seeking work, so there was a considerable turnover. Newcomers spent their first night in a doss-house of sixty-four beds, where they had a bath and their clothes were disinfected.

Most of the military complex was ready for occupation in 1904 and was divided into eight barracks named after the British Army's Indian and Afghan successes: Aliwal, Assaye, Bhurtpore, Candahar, Delhi, Jellalabad, Lucknow and Mooltan. Though much preferred to Bulford 'shanty town', they were not 'the perfection of military housing' predicted by the *Daily Express* in August 1903. Further work was necessary over the next few years, such as, in 1907, heating the large dining-halls and providing bath-houses. In 1906 it was decided, after all, to establish Tidworth as a cavalry station and to spend £25,000 on stables, to be ready by the end of September 1907. The work was finally completed in June 1908 at almost double the estimated cost, with the Scots Greys in the meantime having to stable their horses 4 miles away at Bulford, reducing them to infantry drill.

As a major garrison, Tidworth was well served with facilities, including a theatre, next door to which was the 'Tin Market', built of corrugated iron and containing twenty stalls offering goods and provisions. The Wesleyan Church opened a soldiers' home in 1908, named after Sir Ian Hamilton, the General Officer Commanding Salisbury Plain. The Royal Army Temperance Association had a refreshments room as an alternative to the

Tidworth

Tidworth Barracks. The main soldiers' accommodation is on either side. The tower-like buildings are rooms for NCOs and stores, the cookhouses in the centre are flanked by dining-rooms and the smallest buildings are latrines.

barracks canteens, where beer from Simonds brewery sold at 2d a pint, and the Ram Inn, which was the only civilian pub nearby and often out of bounds after brawls.

All in all, by the standards of early twentieth-century soldiering Tidworth was none too bad a posting. When one unit moved out in 1910, the *Andover Advertiser* reported a soldier's wife as being good enough to declare 'that their quarters were the most decent they had ever been in ... still the majority of the company appeared to be pleased they were going to a place where they could see some hat shops'.

The building of a large ordnance store for ammunition and other materials and served by its own siding started close to the railway station in late 1909. Plans of it made in 1918 show an impressive complex of buildings, store rooms and ancillary shops for motor-lorry repair, woodwork, painting and tinning, as well as a tyre-heating furnace and smithy. A shed for railway engines serving the barracks was built close by, but no one seems to have anticipated the fire risk of such proximity, so a new shed had to be erected further away. Between 5 August and 31 December 1914, 10,000 truck-loads of Government stores were delivered to the depot, which were put at risk in February 1915 by a fire in a hut nearby, which killed a labourer. In his foreword to D. H. Rowland's

For the Duration – The Story of the Thirteenth Battalion The Rifle Brigade, Captain W. B. Maxwell described:

> the mingled profusion and miserliness of the Ordnance Depot at Tidworth. It was like a boot factory in the West End, a severe Harrods, a rather sordid Gamage [like Harrods, a major London department store]. One was made to feel exactly like a shoplifter as soon as one crossed the threshold. Suspicion met one, a blank refusal ushered one out. But if they could not give one what one wanted, they gave one a bit of advice instead – to do without it.

General Sir Henry Beauvoir de Lisle took over command of the 2nd Cavalry Brigade at Tidworth in 1911 and, believing that war was inevitable, instituted a demanding training programme. On 30 July 1914 officers were enjoying a ball at Tidworth when news arrived of heavy fighting in Serbia, causing them to leave early to rejoin their regiments, many of which became part of the British Expeditionary Force. The troopers were delighted, firing off blank cartridges at the thought of real action before busying themselves preparing for France. Any metalwork that might shine was left unpolished, private possessions and spare kit stored (with the latter later being used to equip recruits), and a hundred rounds of ammunition issued to each man.*

There was much speculation among the Officers' Training Corps cadets camped close to the barracks, evidenced by the letters and cards they sent to their families warning that they might be coming home earlier than expected. So it proved, and their places were soon taken by Regulars and Reservists. Very soon, 40,000 men were living in the barracks and under canvas nearby. Among them was the Wessex Division, which had camped on Salisbury Plain on 26 July. Its headquarters moved to Exeter on the 30th, with individual units guarding ports in Devon, Cornwall and Somerset. On 10 August it was back on Salisbury Plain, establishing its headquarters at Tidworth on the 13th before sailing for India.

Several OTC cadets, whose summer camps in the locality had broken up early, also found themselves back there, attached to regiments and awaiting commissions and uniforms. One such was Norton Hughes-Hallett, who wrote a number of letters home from Bhurtpore Barracks, where he was attached to the newly raised 9th Worcestershire Regiment. He was under canvas because there were too many officers for the available barracks beds, but he reported that at the mess one could eat practically what one liked for

---

*Among the units from Tidworth were the 4th Dragoon Guards, whose Ted Thomas was the first member of the BEF to fire at the enemy, on 21 August; 'it seemed to me more like rifle practice on Salisbury Plain,' he said. Minutes before, the 4th's Captain Hornby had sabred a German Uhlan. Also from Tidworth were the 9th Lancers, whose Captain Francis Grenfell won the first gazetted Victoria Cross of the Great War, on 24 August.

only 3s 6d a day. However, he had to write home several times for blankets (for which there was a national appeal in mid-September on behalf of soldiers) and civilian clothes: a dark suit, grey trousers, grey woolly waistcoat, an Old Haileyburian sweater and black shoes. In contrast, in September a recruit from Gloucester, less sartorially needy than Hughes-Hallett, wrote reassuringly home: 'dont trouble to send anything as they have provided us with a shirt, towel & a pair of socks, so I can manage now'.

Cavalry units provided much of the British garrison in the war years, with several Reserve regiments being formed there from Yeomanry formations, many from the west of England. As well as providing drafts for active service, the garrison also had a home defence role. Though it was well sited to respond to an enemy landing on the south coast and an advance on London, any invasion was thought to be more likely on the east coast and in the north. Thus there were contingency plans to transport troops from Salisbury Plain to these areas. Emergency timetables for troop trains were drawn up, with preserved rations and grain to last some 35,000 men three days being stored at Tidworth and Bulford in 1916, together with 12,000,000 rounds of small-arms ammunition and 10,000 rounds for 18-pounders.

Tidworth became the Australian Imperial Force headquarters in Britain in mid-1916, a logical choice with so many local camps occupied by its troops. For most of 1919 staff there were busy coping with the repatriation of their impatient countrymen, with up to 40,000 men at any one time in Wiltshire camps waiting to go home. They were joined in midsummer by 380 men who had been released from the AIF detention barracks at Lewes. Their presence could have done nothing to improve the unsettled air at Tidworth, with remnants of British battalions returning from active service and some British units moving there briefly before being transferred overseas. But compared with elsewhere, troops at Tidworth seem to have been reliable, for some were sent to deal with unrest at Chiseldon Camp in early 1919 and others provided a calming influence during the railway strike in October and the coal strike towards the end of 1920.

In April 1920 Private Cooper of the 9th Lancers escaped from Tidworth guardroom, where he was awaiting trial for desertion. He was mistaken for another man on the run, Percy Toplis, the so-called 'Monocled Mutineer', who was then attracting much newspaper coverage after he had shot a Salisbury taxi-driver at Thruxton Down, 3 miles south-east of Tidworth. After a chase across the Plain Cooper was caught and returned to his cell, where he hanged himself.

The peacetime barracks saw the eventual phasing out of horses for practical purposes in favour of increasing mechanization. The first of the post-war series of the famous Tidworth Tattoo was held in June 1920. 'Torchlight displays' and tattoos had been mounted before the war, notably in June 1906, when, as well as martial themes being presented, there was a re-enactment under searchlights (then a military novelty) of 'Druidical rites' at a replica Stonehenge. Another tattoo, followed by a simulated attack on an armoured train, had been held in August 1906 with Richard Haldane, the Secretary of State for War, present.

TIDWORTH PARK

Tidworth Park was an L-shaped camping-ground (map reference 238474) first used in 1903, south of Tidworth House and straddling the road to Shipton Bellinger. Just inside Hampshire, it was a pleasant site, set in a sheltered, lush valley through which ran the River Bourne and close to Tidworth village. Until garrison churches were built at Tidworth, the Park was used for church parades in the summer.

Before the war the ground was usually allocated to cadet and medical units. In July 1903, 330 members of the Royal Military Academy, Sandhurst, camped there, having marched 15 miles a day from Camberley.

The Park was filled with OTC units from schools in July 1914, but their summer camps broke up early with the declaration of war, leaving the site for men of Kitchener's New Army and the Royal Army Medical Corps.

In 1915 the Royal Engineers' 152nd Field Company was camped alongside the Swindon–Marlborough road at South Tidworth, which suggests Tidworth Park Camp. A. E. Henderson slept with his boots as a pillow, as, with several heavy drinkers in the tent, they could be filled with 'something unpleasant'. Rather than go outside, one man relieved himself in his mess tin and tipped it under the tent wall, leading to a very nasty smell. Henderson was never able to have a bath there. A short time later the 37th Divisional Signalling Company found only latrines on the site, but its commanding officer soon had the place made comfortable, with an improved water supply, baths, wash and cookhouses and dining-rooms – which would have been makeshift structures and tents. Huts were not built there, with the Park continuing as a camping-ground for a variety of troops throughout the war. With the return of peace it resumed its role as a regular venue for cadets.

TIDWORTH PENNINGS

Tidworth Pennings was a camping-ground (map reference 224504) north of Tidworth Barracks and east of Sidbury Hill (off which water sometimes poured on to the site in wet weather). It was first used in June 1903 by the

Royal Military Academy, Woolwich, when twenty-seven officers, 269 cadets and 130 other ranks assembled there.

After the Officers' Training Corps was formed in 1908 cadet camps were held at the Pennings in July and August each year. Bertie Flinderlich wrote home in 1910:

> We were engaged in night operations last night and marched about a bit in the dark. There were no casualties. My torch came in useful. Being the 'white' force we had to tie handkerchiefs round our caps. They were jolly useful in the dark, as our grey is even more invisible than khaki. Today is another field-day.

Accounts of camp life can be found in many contemporary school magazines. The *Carthusian*, for example, reported on an excellent ten days at Tidworth Pennings in 1910. The first two days were spent on battalion drill and preparing attacks. Then the Public Schools Brigade took part in a divisional field day. 'Charterhouse was lucky and saw a great deal of fighting, but its cadets were vastly entertained by a battalion of Territorials who carried waterproof sheets with them to protect themselves from colds,' noted the *Carthusian*. The cadets were inspected twice by the Duke of Connaught, who visited their

These schoolboys of the Officers' Training Corps at Tidworth Pennings have 'curtains' attached to their caps to protect their necks against the sun, a precaution sometimes taken also by Territorial soldiers whose civilian jobs did not accustom them to outdoor life.

camp and took the salute at the march-past of 20,000 troops. Charterhouse tied with Highgate School for first place in a tent-pitching contest and was told by the camp commandant, Colonel Oxley, that its lines were the best: 'Though he himself was a Carthusian, he had to say so because the other officers wished it.'

The Times reported in July 1914 that just under 3,000 public-school boys would be in camp at Tidworth Pennings until 6 August with:

> fairly heavy training in front of them in view of the extreme youth of some of their number ... so far as the camp itself is concerned, their duties are light, for while they have to provide tent orderlies, practically all the fatigue work is performed by Regulars. While in camp, cadets are allowed to have one sovereign pocket-money, which has to be deposited with the commanding officer. This will probably all go in extra provisions, for there is little else on which to spend it in the district.

The camp broke up early because of the outbreak of war, many of the Regulars running it being called away at two hours' notice, but four weeks later public schoolboys, past and present and now potential officers, assembled at the Pennings. About a thousand were there on 1 September, with more expected during the next few days. Among them was H. W. House, who had left Rugby School in July:

> Towards the end of August I had a letter from the officer commanding the Corps at Rugby saying that if I would like to go to a camp for those wanting commissions, I ... must report at Tidworth Pennings. I must go to Rugby, pick up my uniform at the Armoury, get my hair cut, and go on to Tidworth Pennings ... I went on to Tidworth Pennings ... feeling very self-conscious in my Corps uniform ... When we got to the camp, we reported to the H.Q. of the Rugby contingent and were divided into sections according to our houses ... There were a lot of other schools besides Rugby. I think the camp lasted a fortnight, and on the last day we had to report to our Commanding Officers and say whether or not we wanted to apply for commissions.

At home a fortnight later, House received a letter from the War Office saying he had been commissioned into the East Lancashire Regiment and in early October found himself back at Tidworth Pennings joining its 7th Battalion. Its officers shared mess quarters in large marquees with the King's Own (Royal Lancaster) Regiment: 'the food was [provided] by a firm of army contractors [and] was quite good though not elaborate, and the price was moderate. There was plenty of drink obtainable.' House's day began with a half-hour run at 7.30am and finished about 4.30, after which he was free, though expected to dine in mess most nights. At first, conditions were reasonable, but then the weather deteriorated very badly, with the result that the tents, some of which lacked floorboards, were flooded. Eventually the battalion moved into billets in Andover.

Huntly Gordon in *The Unreturning Army* gives a good account of the OTC camp at Tidworth Pennings in 1916:

> we were given demonstrations of every aspect of warfare. We put on gas-masks, hesitatingly entered a tent of chlorine gas and were relieved to find ourselves unaffected by it. We fired Stokes mortars, watching the projectiles shoot up into the air and land with a satisfying crash in a nearby quarry. We worked field telephones, sat in the cockpit of an old aeroplane and generally had an exciting and useful ten days of it.

On 15 August 1917 the New Zealand Rifle Brigade's reserve troops, about 2,000 strong, moved from an overcrowded Sling and established a canvas camp at Tidworth Pennings. At Sling, the soldiers had passed through a succession of specialist instructors, but now this system was modified to place more responsibility on to officers, so as to introduce more *esprit de corps*.

The Official History of the New Zealand Rifle Brigade noted the camp's delightful surroundings and commented that the training grounds were admirable.

> Within the camp itself special attention was directed towards securing the maximum comfort for the men as far as the general conditions permitted. The tents were fitted with wooden floors, and in addition to a liberal issue of blankets each man was provided with a palliasse and bolster. Even hot shower-baths for general use were fitted up in a number of tents. Indeed, as the result of the unceasing labours of Capt. W. E. Christie, than whom there was probably no more efficient quartermaster in the British Forces, the camp at once took its place as the model for the whole of the Southern Command, and General Slater [*sic*: actually Sclater], commanding in the south, on more than one occasion sent representative officers to observe the working of the camp generally, and in particular to note the methods and devices introduced in the quartermaster's branch.

But the Pennings was still a camping-ground for the summer only and on 27 September the Brigade moved to a permanent camp at Brocton. OTC contingents continued to use Tidworth Pennings through the war and beyond. It was never built on.

Near to Tidworth Pennings lies Tidworth Military Cemetery, laid out in 1903. Two of the earliest graves with decipherable inscriptions are those of two children of soldiers, dated 1906. The following year John Charles Rhymes, late of the Royal Welch Fusiliers, and Lance-Sergeant Benjamin Blake of the 1st Royal Fusiliers were laid to rest there. Of the original 423 graves of the Great War period, 100 were of New Zealand soldiers and 159 of Australians, many of who died in the winter of 1918–19 from the flu epidemic.

There are a few Canadian and Royal Flying Corps graves. The remains of eight Germans were removed to Cannock Chase, Staffordshire.

TROWBRIDGE

Trowbridge was originally garrisoned to help quell local riots. Cavalry barracks were built in 1794 at the junction of Bradley Road and Frome Road, with a tall, three-storey block being added in the Victorian period, but by the 1890s they were little used.

When it was decided to send Yeomanry units to the Boer War, Colonel Walter Long (who as a lieutenant had rejuvenated a flagging Wiltshire Yeomanry in 1877 by raising a troop in his home area of Rood Ashton) set up headquarters in the barracks at Trowbridge and quickly formed three companies. Given a choice of inoculation or vaccination against enteric fever (typhoid), the men unanimously chose the former, seen as having fewer side effects.

The Royal Horse Artillery was quartered in the barracks for some years, leaving Trowbridge on 5 August 1914 to go to war. The same day saw a fracas at the station between Reservists and two men thought to be Germans, the fate of the latter frustratingly not being mentioned in the *Wiltshire Times* reports. The barracks were also the headquarters of the 4th Wiltshire Regiment of Territorials and in early August soon became crowded as its members flocked to the town, many being lodged in billets. On 7 September the battalion lined up to leave for Salisbury Plain, presenting a rather motley appearance, with the majority in plain clothes, because supplies of khaki were in short supply. Hardly any men carried rifles and the variety of headgear included straw hats, billy-cocks (bowler hats) and caps. Some of the 4th were delegated to erect tents at Codford for the new Kitchener battalions. The departed soldiers were quickly replaced by recruits forming the 2/4th Wiltshire.

For much of the war the barracks were an artillery cadet school where author and poet Edward Thomas spent several weeks training with the Royal Garrison Artillery towards the end of 1916, living in a tent. He noted:

> it is rather difficult, too, to learn about pulleys and weights and the teaching is almost useless. Partly, too, the very violent physical drill explains it, and a night partly spent in trying to keep rain out of the tent ... we had some walks on Saturday and Sunday, but all the week we are practically confined to barracks as we work till 7.30 and can't go out unless we are in our finery, which is hardly worth while changing into.

WEST DOWN

In 1897 an area east of Tilshead, described by the *Andover Advertiser* as 'an outlandish spot in all conscience', became one of the first parcels of Salisbury Plain to be purchased for military use. Part of it was at first known as Rollestone Artillery Range before taking the name 'West Down' from a relatively small patch of land.

For some years there were two separate camps, not always differentiated between: West Down North (map reference 063494) and West Down South (063481), the former at first being known for a couple of years as 'East Down'. (Confusingly, it was north of West Down.) East Down Camp was used for the first time in 1899; a 290ft well was sunk to provide water and was linked to one at West Down South.

Nearby the open land was ideal for firing ranges. Britain had decided against buying the Vickers Maxim gun, which could fire 1lb shells at the rate of 400 a minute, its noise leading to it being called the 'pom-pom'. With shells costing 6s 6d each, it was too costly for many armies, even with its firing rate reduced to 300 a minute. But during the Boer War the British Army experienced only too well the effectiveness of the weapon, several having been acquired by the Boers. Britain belatedly decided to purchase it and there was much firing practice with it on the West Down ranges.

A plan for the Amesbury branch railway to extend to Shrewton, 3.5 miles from West Down, was cancelled, and the camp remained isolated, though a military road was built from the Bustard Inn past Greenlands Farm and West Down South to Tilshead.

Before the Great War, West Down was one of only three Wiltshire sites, together with Pond Farm and Windmill Hill, that were regarded as suitable for Yeomanry. Yet it and Pond Farm were bleak, remote locations, exposed to the wind, and one wonders why sheltered, more accessible, sites were not chosen.

In October 1914 Canadian troops were based there. In November West Down North was considered the muddiest of the Salisbury Plain camps – hardly suitable for the supply wagons and artillery there. The very wet winter quickly led to conditions becoming atrocious. The Canadians' horses particularly suffered, so a veterinary hospital opened at Keepers Farm, close to West Down South. At this time, West Down South itself is said to have had only two water taps for everyone, including the cookhouse staff. (One presumes water was available directly from the well mentioned above.)

With the departure of the Canadians in February 1915 the West Down sites appear not to have been used until well after the return of peace. Eventually a hutted camp was built west of West Down South, close to the village of Tilshead, and was called Westdown.

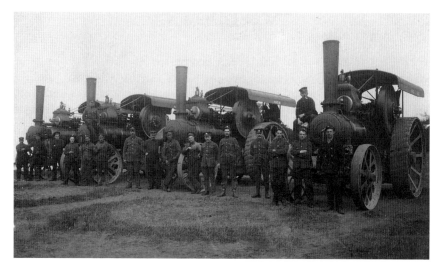

A fine array of traction engines and their Army Service Corps crews at West Down Camp. Their hauling power was crucial in pulling wagons of equipment over the undulating Wiltshire countryside, but their ribbed rear wheels damaged the road surfaces.

WINDMILL HILL

A camping-ground (map reference 253505) was established at Windmill Hill, to the north-east of Tidworth, in 1903. Ever proud of the countryside in its circulation area, the *Andover Advertiser* enthused: 'we doubt if there is another on the Plain that can beat it for picturesqueness of surroundings, as therefrom can be obtained a view of the beautiful woodland scenery that stretches around in the direction of Tidworth Pennings'.

The location was only a mile from Ludgershall Station, a proximity that made the camp popular with troops not eager for an arduous march after a long train journey. An Officers' Training Corps camp was held in July 1914. One cadet, F. P. Roe (later a brigadier), has recalled with shame in *Accidental Soldiers* how, after carrying out a stiff training exercise, which included a 12-mile march, he joined in a song during a rest:

> Why did we join the OTC?
> Why did we join the army?
> Why did we come to Windmill Hill?
> Because we were bloody well barmy.

Windmill Hill

After the cadets had finished their song, their commanding officer, Major C. C. Christie, remarked: 'I hope I have succeeded in training you to be good officers. I have, I see, failed to teach you to be gentlemen. You will return to camp with your rifles at the slope the whole time except for the ten-minute break every hour; there will be no marching at ease.'

With war not yet declared, Major Christie handed out application forms for temporary commissions, each filled in with name and date and merely needing the cadet's signature. He said that war was a certainty and would last at least three years.

The 10th Royal Fusiliers, which at that time recruited from City firms in London, camped at Windmill Hill in early 1915, having spent six weeks in billets in Andover. For C Company its reveille one morning was set for 3.30am, with the men due on parade at 4.15. But the sergeant-major due to wake the men had got 'beastly drunk'. When roused by the sentry, he went back to sleep and did not get up until 4.00, when he was called by the cook-sergeant, who had the men's breakfast ready. Just six officers and the sergeant-major paraded at the time ordered. Eventually the men spent ten hours in 'a heavy and bitterly cold rain' in the 'rottenest' rifle-range butts some had ever seen. The company commander told his men that they had disgraced him and he was no longer proud to command them. Leave was stopped and fourteen men who went to a portable coffee stall without permission were confined to

The Second Volunteer Battalion, Cheshire Regiment, marching from Windmill Hill in 1906.

camp for three days. They protested to the battalion's commanding officer, the NCOs threatened to resign if the company commander did not apologize and knots of sullen men gathered. Higher authority decided that the matter should be dropped and the company commander 'went on leave'.

The 13th Rifle Brigade seems to have had a far happier time when it moved there in April. An exuberant Evelyn Southwell wrote:

> We are having a great time here: there is as far as I can see nothing to complain of except dust, than which I have never seen any thicker in midsummer. It rises in almost solid clouds from a string of thirty or forty motor lorries, such as one meets constantly; while the smallest party of infantry raise enough to make them unpleasant to follow.

Southwell revelled in the training in dry weather in the surrounding countryside, describing a brigade training week as 'the Wonderful Week', and 'one of the happiest I've ever known'. But he was always pleased to sight the two clumps of trees on Windmill Hill when returning from long marches.

From 1915 tents at Windmill Hill were used in the summer to cope with 'overflow' from nearby hutments and Tidworth Barracks. After the war, summer camps continued on the site and a racecourse was laid out. In 1921 it was the only Southern Command camp listed as suitable for Yeomanry and cavalry. Permanent latrines and bathing-pools were added in the post-war years.

OTHER MILITARY BASES

As well as established camping-grounds that were regularly used year after year and hosted hundreds of soldiers in summertime, there were other locations in Wiltshire that occasionally saw smaller camps. Several permanent bivouac sites were laid out for troops to use overnight and before the Great War there could have been few villages in southern Wiltshire that did not have soldiers camping nearby at least once.

Beach's Barn

Beach's Barn (sometimes Beache's or Beech's, 184512), 400yd north of the Fittleton–Everleigh road, was a bivouac camp during the Great War.

Bowood Park

Part of the Bowood Park estate, 2 miles west of Calne, provided an overnight camp for 2,000 men of the 2nd Cavalry Brigade during the September 1903

An ambulance at the Yeomanry camp at Bowood Park in the summer of 1915.

manoeuvres. Newly formed Yeomanry units comprising 2,500 men from Wiltshire, Dorset, Somerset and Hampshire spent the summer of 1915 there, with the Army Service Corps unloading goods at Black Dog siding on the Chippenham–Calne branch railway. The park was owned by the Marquis of Lansdowne, Honorary Colonel of the Royal Wiltshire Yeomanry, who allowed water to be pumped from his 10,000gal reservoir at Lady Spout, 1.5 miles away. A hospital was established in the servants' quarters at Bowood House, though when two yeomen riding a motorcycle collided at Sandy Lane with a car driven by Sir Audley Neeld, President of the Wiltshire Agriculture Association, they were taken to Calne Hospital.

Brigmerston

Brigmerston, in the Avon Valley, had a bivouac site (156459) near to Knighton Copse, very close to one of the bridges built by the Army over the River Avon so that troops could move between the central and eastern parts of Salisbury Plain.

Burderop Park

Burderop Park, a mile west of Chisledon, was owned by T. C. P. Calley who frequently made his land available to the Wiltshire Yeomanry for camps before the Great War. In 1904, 564 of all ranks were on parade there, out of a total strength of 602, a noteworthy turnout. (Occasionally Yeomanry units visited other country estates for their camps and exercises, Pyt House at Tisbury and Charlton Park near Malmesbury both being used in this way.)

Butler's Cross

Butler's Cross (023498), near Market Lavington, was the base for a Royal Engineers Meteorological Section, whose headquarters were at Stonehenge Airfield. There was probably a handful of huts there – six can be seen in an aerial photograph of 1924.

Chapperton Down

The Chapperton Down Artillery School was established between Imber and Tilshead in 1916, with its headquarters being 15 miles away in Salisbury. Though its ranges had a telephone system and splinter-proof observation stations, there was no living accommodation. Officers and their batmen were billeted as far away as Salisbury and in Tilshead and Imber Court, in whose attic at one time more than forty soldiers slept, with six officers more comfortably housed. The proximity of the artillery school meant that the Imber villagers and their houses regularly suffered noise and concussion; on some days they were unable to travel to nearby communities because roads were under fire.

Fittleton

Situated in the Avon Valley, Fittleton was a pre-war bivouac camping-ground (143500), south-west of the hamlet of Fifield. It was close to a crossing established by the Army over the River Avon.

Ford

A tented camp at Ford (164330), a mile east of Old Sarum, was an assembly point for troops for the 1898 manoeuvres and was used thereafter for overnight halts.

Greenlands Farm

Greenlands Farm (sometimes Greenland, 067472) was a remote spot 5 miles from Lark Hill and 2 miles from Tilshead on the western edge of the Lark Hill artillery ranges. (It should not be confused with Greenland Farm, 3 miles to the south-east and a half-mile south of Rollestone Camp.) The farm was taken over by the Army and became a base for men responsible for maintaining the range and ensuring its safe use. W. A. Wilson of the Canadian Field Artillery, who was based there in the early autumn of 1914, noted that the 'gun park was on a very exposed hill, bitterly cold in January'.

There was a serious accident close to the farm in 1920 during experiments to construct a tank target that would travel at 20mph. A wooden frame secured to a corrugated-iron sheet was attached to a steam lorry. This primitive arrangement proved difficult to get under way, but after several attempts it shot off at 40mph, jumped from its course on the crest of a hill and killed a soldier.

By 1926 the original farmhouse had been demolished and Greenland Camp, as it had become known, consisted of a few original farm outhouses and a collection of huts and brick buildings.

Little Somerford

Quarters at Little Somerford were grandly described as 'barracks' in September 1916 by a deputy medical officer of health who said they should be condemned; ten men were living in a small railway carriage suitable only for four. Though the precise site is unknown, they were guarding the bridge taking the GWR's Swindon–Bristol line over the River Avon. No doubt billets and huts were provided for soldiers guarding other key installations: a list of 1917 'military occupancies' notes under 'Wiltshire' such huts close to Monkton Viaduct (which was just over the county border, 3 miles south-east of Bath).

New Copse Farm

New Copse Farm (027504), between Tilshead and Lavington, was used before the war as quarters for soldiers administering the firing ranges and in 1914–15 by the Canadians' First Field Artillery.

Newfoundland Farm

Two miles north-west of Lark Hill (107482), Newfoundland Farm also housed range parties and equipment.

Patney & Chirton Station

A field near Patney & Chirton Station was used as a rest site for troops arriving by rail before they marched to camps on the Plain.

Ratfyn

Ratfyn (153423) was close to where the Lark Hill Military Railway crossed the River Avon. On the west side of the railway viaduct were three reception sidings – actually loops – to accommodate wagons carrying stores and materials for Lark Hill Camp; on the eastern side was a locomotive shed and electricity station.

Ridge Quarry

Ridge Quarry, south of Corsham (873680), was taken over from the Bath Stone Company by the Government in 1915 for the storage of TNT and cordite, with nearby public footpaths being closed. The following year, the workings were inspected by two senior officers, who were so impressed by the company's general manager, Sturge Cotterell, that he was appointed Superintendent of Munitions Stores.

The quarry offered 12 acres of storage space, half of which was cleared and converted during the Great War. A tramway connected the mine with the Great Western Railway main line at Corsham Station, 1.5 miles away. The depot was abandoned shortly after the war, but in 1936 the War Department purchased Ridge and two nearby quarries for £47,000, converting them for ammunition storage. This was the start of some very ambitious work whereby fifteen quarries in the area were turned into underground depots.

Shepherds Shore

Shepherds Shore (045662), between Devizes and Beckhampton, provided an occasional camping-ground, which was used by the Western Counties Infantry (Volunteer) Brigade in 1891 and the Wiltshire Yeomanry before the war, and by the Army Service Corps in August 1915.

Silk Hill

Situated between Milston and Tidworth, Silk Hill was the site of an occasional bivouac camp (187468) during the Great War.

Stockton

Stockton, in the Wylye Valley, saw large concentrations of recruits in the first months of the war, among them those of the 10th Devonshire Regiment who arrived on 26 September 1914 to live under canvas. Each man was issued with a knife and fork, but no spoon. The tent orderlies had difficulty in serving dinner from dixies until ladles arrived from Salisbury. Officers and NCOs took drill books on parade to help them give instructions and attempted some very limited field exercises. As winter drew in, the 10th moved into billets in Bath and then, in April 1915, to new huts at Sutton Veny. Numbers 1 to 4 of the Codford camps were built to the north of the village. Australian troops played cricket matches at Stockton House.

Upton Lovel

A tented camp was laid out near Upton Lovel shortly after the outbreak of war as part of the concentration of recruits in the Codford locality. Army Service Corps men of the 29th Division were so glad to leave in February 1915 that they wrote a lengthy song which included the following sentiment:

> The Colonel wrote a letter to our Captain t'other day,
> Saying that the Wessex boys would soon be going away,
> The thought of leaving Upton Lovel filled them with delight,
> The lads became excited and they yelled with all their might.

Chorus:

> It's not far up to Leamington,
> Where we are to go,
> It's not far up to Leamington,
> And the 29th we know,
> So it's good-bye Upton Lovel,
> Farewell Codford in the wet,
> For now there's hope of going to Belgium
> So we're not down-hearted yet.

(Though the 29th Division was formed in and around Leamington between January and March 1915, its Army Service Corps divisional train – responsible for transporting baggage – comprised men from the Wessex Division.)

Warminster

Fair Field in Warminster was taken over in 1914 for a military rail depot. 'North' and 'South' camps provided huts for upwards of 1,400 men.

Wilcot

Situated a mile north-west of Pewsey, Wilcot was used for overnight bivouacking during the Great War. The village was a pleasant one and the site would have been equally agreeable, but for the fact that for the soldier the only shelter was a blanket stretched over pegs. When the 13th Rifle Brigade stayed there in 1915, its food was late arriving and Lieutenant Evelyn Southwell had to procure some biscuits for his men. He himself stayed in the vicarage, where he enjoyed a 'large breakfast'.

Wilton

Early in 1915 at Fair Field, Wilton new railway sidings were laid to help with the unloading of Government stores, probably intended for the Army camp being built at Fovant, 5 miles away. Army Service Corps units were based there to unload and transport the stores. The next year, men of the 650th Mechanical Transport Company were billeted with local villagers, spending the summer under canvas in Wilton Park, before returning to billets in October, just as work was starting on building a small hutment at Fair Field. The huts were ready for occupation on 17 December and included a sergeants' mess, corporals' room, regimental institute (providing off-duty facilities), ablution hut, bath-house, cookhouse, latrines and guardroom. The old Wool Loft in Wilton's Market Square was turned into a workshop and stores, while Randall & Pretty's garage in North Street became a machine shop that made spare parts more quickly than they could be obtained from manufacturers. Several fitters among the soldiers sent home for their own tools.

* * *

During the war there were small camps for detachments of soldiers in many parts of the county, as well as those of the weekend type for members of the Volunteer Training Corps, National Guard and so on. In July 1917 wet weather forced cadets from forty-seven secondary schools camped on Marlborough Common (where the 258th Company, Army Service Corps, had lived in their lorries in the summer of 1915) to go into billets at Marlborough College.

Appendix: Fovant Military Badges

In the summer of 1916 men of the London Rifle Brigade carved a massive regimental badge in the steep chalk escarpment above their camp at Fovant. The hours they could spend on the task were limited because the hillside was above a much-used rifle range, so a great deal of the work was done from dawn until 7am. Other badges carved nearby in the same period and preserved after the war include those of the 6th City of London Rifles, the Post Office Rifles, the Devonshire Regiment, the YMCA, the Australian Imperial Force and, at Sutton Mandeville, the Royal Warwickshire Regiment and the 7th City of London Battalion. (These last two survived until the early twenty-first century, when the cost of upkeep became excessive.)

Though many have applauded the labours of the men who did all the manual work, few have paid tribute to the skills and patience of those who laid out the outlines on a steep hillside. That of the London Rifle Brigade is said to have been done by an architect. Many units used grid patterns to mark out

After the Great War the smaller badges shown here – some little more than scratchings – were allowed to fade away, leading to an overall improvement in the appearance of the hillside displays.

Appendix: Fovant Military Badges

their badges and semaphore signals to make corrections – all useful training. Some badges were constructed by digging trenches in the turf and filling them with chalk; others merely consisted of chalk placed on the grass, which soon washed away.

Volunteers from Hurdcott Camp created an outline map of Australia at Compton Chamberlayne, together with other Australian badges. Mrs F. A. Cranstoun, a Tasmanian Red Cross worker visiting the area in 1918, wrote home that the owner of the land:

> had sued the Commonwealth [of Australia] for damages. He was offered £1500, but refused it, went to law, lost the case, and had to pay his own costs. Truly a just punishment. He should have been well pleased to have his entirely useless hill turned into a work of art for all time.

The figure seems high and it has not been possible to verify the story.

W. H. A. Groom commented 'I am not sure that the large regimental badges cut in the chalk hillside with much hard labour was an improvement on nature, but I suppose it denoted a praiseworthy effort of 'Anything you can do we can do better' between battalions and helped to maintain *esprit de corps*'.

Badges of the Great War period no longer visible include those of the Royal Army Service Corps, Royal Army Medical Corps (RAMC), Machine Gun Corps and Queen Victoria's Rifles – and the outline of a kangaroo. The Queen Victoria's Rifles spent three Sunday afternoons of 'compulsory voluntary fatigues' on theirs, using picks and axes – and fire buckets from the huts. Local newspapers in 1919 reported that members of the Voluntary Aid Detachment had also carved their badge. Some soldiers are said to have made a Red Cross, possibly of brickbats or broken tiles, to the fury of the hospital matron. This may have been an embellishment of the existing RAMC badge, which included a cross.

Some unit titles (as distinct from badges) were also cut, but the immediate post-war effect was an untidy one and many smaller designs and some larger ones were allowed to become overgrown and disappear. Titles no longer to be seen include '9th Royal Berks', '35th TR' and '37th T.R.'. The 9th Royal Berkshire Regiment became the 37th Training Reserve Battalion on 1 September 1916 and was based at the Royal Engineers' camp at Fovant in March 1917; it was dissolved in May 1917 after routine service in France handling stores. The 35th Training Reserve Battalion evolved out of the 7th Dorset Regiment and was based at West Farm in March 1917. The Drums section of the 35th Training Reserve Battalion also carved its unit title and an outline drum above East Farm; the effect was elegant but faint, the design being squeezed in-between two bolder badges, and it was allowed to disappear.

Appendix: Fovant Military Badges

Other badges were carved after the Second World War. The remaining carvings are maintained by the Fovant Badges Society, which has produced booklets on the camps and badges and each year holds a memorial service for the men who were based locally.

Other military hill carvings in Wiltshire include the 'Bulford Kiwi' cut above Sling Camp in 1919 and the Australian Imperial Force badge outside Codford. Both can still be seen. A postcard of Sutton Veny shows an indistinct – and presumably short-lived – badge on a hillside outside the village. Some units also laid out their badges using whitewashed stones on the ground in their hutments, though with the rapid turnover of troops these would have been temporary.

Selected Bibliography

Microfilm copies of most local newspapers are available at Wiltshire and Swindon History Centre, Chippenham, with some other Wiltshire libraries holding those appropriate to their locality. Hampshire's *Andover Advertiser* had two regular columns, one of news from Tidworth, and the other entitled 'Salisbury Plain Notes', describing activities on the eastern Plain and announcing local military appointments and Army instructions. *The Times* provides many details of summer camps and exercises in Wiltshire. In the cases of the larger manoeuvres, some of its descriptions of troop movements are very precise, often naming localities that would only have been known to local people.

Bailey, David, *The Story of Chiseldon Camp*, part I 1914–1922 (Chiseldon Local History Group, 1998).

Bavin, W. D., *Swindon's War Record* (Swindon, 1921).

Bridgeman, Brian and Barnsley, Mike, *The Midland & South Western Junction Railway* (Alan Sutton, 1994); includes photographs of military activities at Chiseldon, Ludgershall and Tidworth.

Bruce, J. M., *The Aeroplanes of the Royal Flying Corps (Military Wing)* (Putnam, 1982); describes many aircraft that flew on Salisbury Plain before the Great War, including entries for the 1912 Military Aeroplane Competition.

Buckton, Henry, *Salisbury Plain* (Phillimore, 2008).

Campbell, Len, *1st Canadian Contingent on Salisbury Plain* (Amesbury, 1996).

Carter, G. B., *Porton Down* (HMSO, 1992).

Clarkson, Gresley, *A Very Man: Donald Drummond Clarkson 1880–1918* (Access Press, 2005); letters from an Australian at Fovant.

Corden, Chris, *The Plain* (Halsgrove, 1998); published in conjunction with a four-part television series transmitted in 1998; reproduces many postcards and other archive material.

Daniels, Peter, & Sawyer, Rex, *Salisbury Plain* (Chalford, 1996); reproduces many photographs and postcards of early military training.

Davies, Mavies, *The Beginning in Peace* (Salisbury, 1991); *Valiant & Determined* (Salisbury, 2006); the Red Cross in Salisbury between 1912 and 1919.

Drew, Lieutenant H. T. B., *The War Effort of New Zealand*, vol 4, (Whitcombe and Tombs, 1923), pp248–53 and 266–70; describes Sling and Codford camps.

Duguid, Colonel A. Fortescue, 'On Salisbury Plain', *Official History of the Canadian Forces in the Great War*, vol 1 and, published separately, appendices and maps (J. O. Patenaude, 1938); the appendices contains letters and telegrams to and from General Alderson while on Salisbury Plain.

Selected Bibliography

Dunalley, Baron (Henry Prittie), *Khaki and Rifle Green* (Hutchinson, 1940); pre-war training on Salisbury Plain and demobilization at Chisledon.

Gill, E. W. B., *War, Wireless and Wangles* (Blackwell, 1934); Devizes Wireless Station.

Grayer, Jeffery, *Rails Across the Plain, The Amesbury & Bulford Branch, The Larkhill Military Railway* (Noodle, 2011).

Groom W. H. A., *Poor Bloody Infantry* (William Kimber, 1976); life at Fovant Camp.

Hennessey, C. F., typescript memoirs, Imperial War Museum, London; life at Sand Hill Camp.

Hill, Anthony, *Young Digger* (Penguin, 2002); the story of 'Digger', the boy mascot at Hurdcott.

Howell, Danny, *The Wylye Valley in Old Photographs* (Alan Sutton, 1988); *Remembering the Wylye Valley* (Danny Howell, 1989); these and other books by Mr Howell reproduce many old postcards and photographs of early military activity and include personal accounts of life in villages close to camps in the Warminster area.

Howson, H. E. R., *Two Men* (Oxford University Press, 1919); includes letters written by Evelyn Southall at Windmill Hill in 1915.

Hughes, Major C. W., *The Forgotten Army* (unpublished typescript, Imperial War Museum); training at Marlborough and Sutton Veny.

Imperial War Graves Commission, *The War Graves of the British Empire: Cemeteries and Churchyards in the County of Wiltshire* (1930).

James, N. D. G., *Gunners at Larkhill* (Gresham, 1983); also covers camps at Durrington, Fargo, Hamilton, Greenlands Farm and Rollestone, the Lark Hill Military Railway, and flying at Lark Hill.

James, N. D. G., *Plain Soldiering* (Hobnob, 1987); camps and barracks on Salisbury Plain.

Lighthall, W. S., memoirs, Imperial War Museum, London; Canadians 1914–15.

Maclean, J. Kennedy and T. Wilkinson Riddle *The YMCA with the Colours* (Marshall Brothers, 1915); the YMCA on Salisbury Plain in late 1914.

Maggs, Colin, *Branch Lines of Wiltshire* (Alan Sutton, 1992), pp120–7 and 140–6.

McCamley, N. J., *Secret Underground Cities* (Leo Cooper, 1998); quarries near Corsham.

McCracken, Mrs G. M., *Looking Back on Sutton Veny* (1981).

McKeon, W. J., *The Fruitful Years* (Wellington, not dated); training of New Zealanders at Sling.

McOmish, David, Field, David and Brown, Graham, *The Field Archaeology of the Salisbury Plain Training Area* (English Heritage, 2002).

Morice, Janet, editor, *Six-Bob-a-Day Tourist* (Penguin Books, 1985); an Australian at Wiltshire camps.

Norris K. P., *The Utilization of Land by the Military in Winterbourne Dauntsey and Winterbourne Gunner 1917–1990* and *The Development of the Porton Campus* (Wiltshire & Swindon History Centre 2916/8 and 2916/12).

Selected Bibliography

Powell, Anne, editor, *A Deep Cry* (Sutton, 1993); Great War poets.
Priddle, Rod, *Wings over Wiltshire* (ALD Design & Print, 2003); Wiltshire airfields.
Rothstein, Andrew, *The Soldiers' Strikes of 1919* (Macmillan, 1980); unrest at Lark Hill and on Salisbury Plain airfields.
Siepmann, Harry, *Echo of the Guns* (Hale, 1987); Christmas 1915 at Heytesbury.
Siggers, Dr John, *Wiltshire and its Postmarks* (1981).
Street, A. G., *The Gentlemen of the Party* (Faber & Faber, 1936); fictionalized account of life in the Fovant area during the Great War.
Taylor, F. A. J., *The Bottom of the Barrel* (Regency Press, London 1978); Chisledon Camp.
Taylor, John W. R., *Birthplace of Air Power* (Putnam, 1958); history of CFS Upavon.
van Emden, Richard, editor, *Tickled to Death to Go* (Spellmount, 1996); memoirs of Benjamin Clouting, 4th Dragoons, at Tidworth Barracks from December 1913 to August 1914.
Wheatley, Dennis, *Officer and Temporary Gentleman 1914–1919* (Hutchinson, 1978); 2/1st City of London Royal Field Artillery at Heytesbury Camp and near Imber.
Wyeth, Romy, *Warriors for the Working Day – Codford during Two World Wars* (Hobnob, 2002).

Miscellaneous

Manor Farm, Codford St Peter, Duplicate Letter Book (University of Reading Library: Microfilm P388).
Greenhill House, Sutton-Veny (c1917); souvenir brochure of Australian YMCA facilities, reproduced in *Warminster and District Archive*, issue 4, May 1990.
The History of the 1st Volunteer Battalion (Duke of Edinburgh's) the Wiltshire Regiment (Bennett Bros, 1919).
Snap Shots of the 15th Battalion The Prince of Wales's Own (West Yorkshire Regiment) (Richard Jackson, 1917); includes forty-seven photographs of Fovant Camp in its early days.
Ordnance Survey 1:63,360 Popular Edition map, 'Salisbury Plain', in colour, 1920, shows much wartime military infrastructure.
Plan by Lieutenant Bruce RAMC and Clarence C. Hancock (surveyor, Warminster Rural District Council), made in May 1915, superimposing the outlines of hutted camps in the Wylye Valley on a 1:10,560 map (Wiltshire and Swindon History Centre, Chippenham: G12/700/IPC).
Plan of wireless station near Devizes, 1915 (National Archives: WO 78/4326).
R. C. C. Clay collection, Salisbury and South Wiltshire Museum; includes Royal Engineers' 1:10,560 map dated 8 February 1916, showing Fovant, Hurdcott and Sutton Mandeville Camps, 1:5,280 plan of the Fovant Estate dated 1919, showing Fovant Camp, photographs taken at the camp hospital, and a register of some 530 men employed on building Fovant Camp.

Index

*Page numbers in **bold** refer to illustrations*

Adlam, Lt Tom 167
airfields
  Boscombe Down 95
  Lake Down 95, 97, 130
  Lark Hill 19, 25, 88, 90, 92, 98, 101
  Lopcombe Corner 95
  Manningford Bohune 95
  Market Lavington 95
  Netheravon 24, 25, 77, 92, **93**, 98, 101
  Old Sarum 95–6, 172
  Stonehenge 96, 161, 171–3, **172**
  Tilshead 96
  Upavon 25, 92–5, 98, 154
  Yatesbury 88, 96, **97**, 104, 106, 149
alcohol & drunkenness 59–60, 63, 135–7, 178, 243
Aldershot 11, 16
Alderson, Lt-Gen. Edwin 59, 65, 136, 182
Alexander, C. M. 131
Allen, Clifford 218
Allen, Gnr John 118
Amesbury 63, 100, 130
Andover 15, 119, 171, 219
Andrews, Hubert 116
Antrobus, Sir Cosmo 103, 171
Appleshaw 130
Ashwell, Lena 137, **138**
Asser, Verney 111
Atwood, Rev. George 67
Australian troops 26, 41, 42–3, 44, 46, 56, 57, **57**, 66–70, **87**, 111, 131, 132, 135, 139–40, 146, 151, 161, 166, 168, 181, 187, 192, 201, 202, 204, 205, 209–11, **209**, 213, 218, 221, 228, 231, 239–41, 245, 249, 259
Avon India Rubber Company 117
Aylwin, Francis & huts 206, **206**

Bader, Pte George 100
banks 191
Barber, Horatio 88
Barclay, James B. 201
Barker, Mabel 144
Barnes, Sgt R. S. **165**
Bates, John 186, 229
Baxter, Archibald 72
'Beaumont', André 90
Beer, Cpl William 68, 179–80
Belgians 121, 132, **132**
Bellomo, Signor 107

Bemerton 171
Benett-Stanford, John 135–6
billeting 19, 118–20
Bishopstrow 67, 146
Blake, Lce-Sgt Benjamin 249
Blenkarne, Sgt-Maj. Percy 73
Boas, Harold **240**
bombs 35–6, 186, 208, 214
Bottomley, Horatio 164
Bracknell Croft 196, 197
Bradford-on-Avon 31, 118, 134, 170
Bratton 124, 226
Brimstone Bottom 242
Briscoe, Lt Ross 34–5
Bristol Flying School 73–4, 80, 89
British & Colonial Aircraft Co. 88–9, 92, 212
British Army units
  corps
  Army Cyclist 187
  Army Service 54, 68, 79n, 178, 255, 258, 259, 262
    10th 202
    12th Horse Transport 24
    258th 260
    348th 88
    499th MT 197
    650th MT 87–8, 260
  Non-Combatant 199–200
  Royal Army Medical 246, 262
  Royal Army Ordnance 221
  Royal Engineers 12, 13, 16, 90, 92, 143, 161, 164, 195, 197, 198, 216, 223, 241, 256
  Royal Flying 90, 182, 213, 220
  Women's Army Auxiliary 180, 181, 216, 226
  artillery
  Honourable Artillery Company **17**, 200
  Royal Artillery 157, 213
  Royal Field Artillery 136, 157, 199
  Royal Garrison Artillery 147, 250
  Royal Horse Artillery 250
    115th Brigade 118
    169th Brigade 207
    291st Brigade 208
    2/1st City of London 52
    10th City of London 84
  brigades
  2nd Cavalry 244, 254
  2nd London 78
  4th Cavalry 12, 177
  6th Light 18

  55th Infantry 191
  58th Infantry 216
  Public Schools 247
  South West Infantry 233
  Western Counties Infantry 258
  divisions
  1st (London) 18
  2nd (London) 18
  18th 41, 191, 207
  19th 41
  20th (Light) 41
  25th 188
  26th 41, 176
  29th 259
  34th 229, 239 & n
  37th 41–2
  58th (London) 103, 183
  60th (London) 79, 299
  96th 193
  South Midland 18
  regiments
  Argyll & Sutherland Highlanders 129
    10th 186
    11th 42
  Bedfordshire Regiment, 7th 167
  Black Watch 60
  Cheshire Regiment, 8th 184, 216
    10th 188, 189
    11th 188, 189
    13th **29**, 100
  City of London Yeomanry 222
  Coldstream Guards 22
  Denbighshire Hussars Imperial Yeomanry 178
  Devonshire Regiment 28, 261
    10th 28, 186, 189, 258
    14th 224
  Dorset Regiment, 1st 224
    7th 262
  Dragoon Guards, 1st 12
    3rd 12
    4th 244n
  Duke of Cornwall's Light Infantry 189
    5th 20
  Duke of Wellington's (West Riding) Regiment, 2nd 137
  Durham Light Infantry, 17th 45
  East Kent Regiment **165**
  East Lancashire Regiment, 7th 119, 220, 248
  East Yorkshire Regiment, 11th (2nd Hull) 35
  Essex Regiment, 10th 191

Index

British Army units (*continued*)
  Gloucestershire Regiment 15
    2/5th 43, 157, 217
    15th 186, 187
    16th 186
  Grenadier Guards, 4th 156
  Hampshire Regiment 24
    16th 186
  Hussars, 8th 18, 20
    12th 178
    18th 25, 136
  'King's Colonials' 222
  King's Own (Royal Lancaster)
    Regiment 248
  King's Royal Rifle Corps, 10th,
    206
  King's Shropshire Light Infantry,
    7th 189, 190
  Lancashire Regiment, 7th 220
    10th 157
  Lancers, 9th 245
  Life Guards, 1st 184
  Lincolnshire Regiment, 7th 220
    10th 221, 237
  London Regiment
    2/2nd **140**
    4/2nd 208
    4th 208
    5th (London Rifle Brigade)
      203, 204, 261
    6th 139, 261
    7th 236, 261
    8th 261
    9th 261
    2/9th 231
    10th 22
    14th 18, 24
    2/14th 230
    2/15th 45, 130, 141, 229, 230
    2/16th 230
    2/18th 36, 188, 231
    3/19th 110–1
  Lothian & Border Horse
    Yeomanry, 1st 207
  Loyal North Lancashire Regiment
    174
    8th 188
    9th 188
  Manchester Regiment, 3/8th 192
  Monmouthshire Engineer Militia
    178
  Northumberland Fusiliers, 19th
    119, 229
  Oxfordshire & Buckinghamshire
    Light Infantry **15**, 203, 229
    7th 232
    8th 176, 203
  Rifle Brigade, 12th 119
    13th 126, 244, 254, 260
  Royal Berkshire Regiment, 9th
    262
  Royal Fusiliers, 1st 249
    10th 253–4

  23rd 35
  25th 138
  Royal Inniskilling Fusiliers 173
  Royal Irish Regiment 15
  Royal Scots, 15th 36
  Royal Scots Greys 12, 176
  Royal Warwickshire Regiment
    1st 116
    13th 236, 261
    14th 45
  Royal Welch Fusiliers 249
    8th 184
  Royal Wiltshire Yeomanry 22–3,
    23n, 28, 118, 250, 258
  South Wales Borderers, 4th 184,
    217
    5th 234
    11th 206
  Suffolk Regiment 9th 157
  Welsh Regiment, 8th 184
    10th 189
  West Yorkshire Regiment, 3rd 37,
    103, 162, 200
  Wiltshire Regiment 108, **109**
    1st 22
    1st V. B. 124
    2nd V. B. 147
    4th 24, 102, 188, 250
    2/4th 250
    5th 195
    6th 220
    7th 29–31, 38, **39**, 48–9, 99,
      158, 195, 237
  Worcestershire Regiment, 9th
    191, 244
    10th 220
  York & Lancaster Regiment,
    12th 157
  Training Reserve Battalions
    33rd 236
    35th 262
    37th 262
    45th 55
British Expeditionary Force 23, 25
Brocklebank, Canon 228, 231
Brooks, Supt 143
Brown, Frank S. 160
Bruce, J. M. 90
Bruckmann, Pte L. 108
Buchanan, Supt 146
Burchardt-Ashton, Lt A. E. 90
burial grounds and graves 62, 108,
  167–8, **168**, 194, 202, 205,
  249–50
Button, Cpl Arthur 36

Calley, Col T. C. P. 184, 255
Calne 107
Calstone reservoir 99
camp and barracks building 47–9,
  49, 50–2, 52–3, 61, **132**
camps and barracks
  Beach's Barn 254

Bowood Park 254, **255**
Boyton 176–7
Brigmerston 255
Bulford 12, 24, 34, **42**, 47, 68,
  104, 105, 126–7, **148**, 174,
  177–81, **179**
Burderop Park 184, 255
Bustard 59, 60, 61, 137, 174,
  181–3, **182**
Butler's Cross 256
Chapperton Down 81, 173, 256
Chisledon 34, 54, 73, 78, 104,
  116
Choulston 92
Codford 36, 45, 49–51, 53, **69**,
  70, 71, 99, 100, 102, 104,
  115, 122, 131, 132, 133,
  136, 141, 149, 151, 157,
  188–94
Corton 83, 169, 176–7, **177**
Devizes 31, 116, 121, 157
  wireless station 197–8, **198**
Draycot 54, 82, 184
Durrington 34, 48, 83, 102,
  103, 162
East Down 15
Fargo **51**, 85, 174, 200–202, **201**
Fittleton 256
Ford 256
Fovant 37, 43, 45, 50, 55, 56–7,
  70, 104, 138, 102, **163**, 171,
  202–5
Greenlands Farm 256–7
Hamilton 20, 24, 48, 205–7
Heytesbury 83, 100, **129**, 131,
  159, 207–8
Hurdcott 35, 36, 46, 56, 57, 69,
  139, 145, 157, 202, 208–11
Lark Hill 12, 48, 53, 61, 64, 74,
  104, 128, 140, 161, 169,
  170, 173, 211–5, **212**
Little Somerford 257
Longbridge Deverill, see Sand Hill
Marlborough Common 260
Netheravon Cavalry School 92,
  216
New Copse Farm 257
Newfoundland Farm 257
Park House 15, 47, 131, 141,
  157, 169, 216–8, **217**
Patney & Chirton 257
Perham Down 15, 22, 47, 55, 66,
  67, 73, 131, 135, 142, 151,
  157, 158, 173, 219–21, 239
Pond Farm 59, 60, 61, 77, 222–3
Porton 74, 82, 122, 151, 223–7
Ratfyn 54, 80, 258
Rollestone 48, 64, 116, 147, 169,
  201, 228
Sand Hill 33, 45, 82, 140, 141,
  228–30, **230**
Shepherds Shore 197, 258
Sherrington 99, 232–3, **232**

269

Index

Silk Hill 258
Sling 8, 24, 35, 44–5, 47, 55, 61, 71, 72, 77, 146, 158–9, 161
Stockton 28, 258–9
Sutton Mandeville 202, 236
Sutton Veny 31, 36, 48–9, 55, 56, 67, 103, 104, 111, 129, 131, 132, 138, **140**, 144, 157, 158, 206, 237–41
Tidworth 12, 24, 26, 34, 35, 66,99, 127, 137, **138**, 140, 241–6, **243**
Tidworth Park 20, 173, 245
Tidworth Pennings 20, 157, 173, 246–9, **247**
Trowbridge 24, 157, 250
Upton Lovel 169, 259
Warminster 259
West Down 15, 24, 59, 60, 77, 99, 102, 131, 159, 251, **252**
Wilcot 42, 259
Wilton 260
Windmill Hill 20, 38, 158, 252–4, **253**
Canadian soldiers 26, 37, 45, 47, 59–66, **63**, 81, 99, 116, 136, 139, 159–60, 168, 182, 187, 212, 222, 251, 257
Carey, Lt Cyril 35
censorship 101–4, 149, 162
Challenger, G. H. 89
Charlton Park 132, 134, 256
Charterhouse School 247–8
Childers, Erskine 90
Chinese labourers 121
Chippenham 28, 104, 118, 124
Chitterne 101
Chivers, W. E. 114–5, 184, 224
Chorlton, Arthur 178
Christie, Capt. W. E. 249
Christie, Maj. C. C. 253
Chubb, Sir Cecil 171
churches and services 126, **127**, 127–8, 192
cinemas and theatres 137–40, 190, 202, 217
Clark, Capt A. M. 73
Clarkson, Donald 56–7, 69, 139, 152, 153, 158
Clouting, Ben 24
Cockburn, George 88
Codford 74, 136, 145
Cody, Samuel F. 88, 90, **91**
Cole, Dr Sydney 135
Cole, Sgt 107
Colebourn, Lt Harry 64, **65**
Collins, A. H. and E. K. 121
Combes, Bob 68
Combes, Daniel 55, 204
Combes, John 121
Compton Chamberlayne 70, 262
Connaught, Duke of 44–5, 176, 219, 221, 247

Conneau, Lt 90
conscientious objectors 72, 218, 236
Cooper, Pte 245
Cotterell, Sturge 258
Cranstoun, Mrs F. A. 262
Crawford, Frederick 213
Crockerton 101, 231
Cruickshank, Sub-Lt W. T. 105

Dauntsey Agricultural School 61, 167
de Candole, Alec 155
de Lisle, Col Sir Henry Beauvoir 26, 244
demobilization 72–3, 82, 161–6, 214
Devizes 61, 104, 123, 124, 143
  lunatic asylum 135
  prison 110
de Winckel, Cpt Ignio 91
Dickeson & Co 114, **115**
Dickson, Capt. Bertram 89
Dinton 166
disposal of war surplus 169–73
Dodsdown 76
Duffell, W. J. 210
Dunne, Gen. John Hart 108
Durkin, Joseph 111
Durrington Walls 199

Elkins, Wilfred 116
Enford 85
Enham 168 & n, 171
entertainers 137–8

farming **21**, 120–2, 124–5
Fedarb, Frederick 129
Fenwick, Robert 91
Field, Henry 156
Fighledean 95
Fisherton de la Mere 191
Flinderlich, Bertie 247
food 54–8, 61, 180, 192, 210, 236, 248
Fort, Pte 202
Fosbury 119
Francis, George 227
French, F. M. Sir John 19, 43
Fritsch, Otto 100
Fuller, J. T. 148
Fulton, Capt. John 88

Gardiner, Henry 54
Gardiner, Tom 231
gas 34, 52, 216, 224–5, 249
George V 41, **42**, **43**, 60, 64, 66, 80, 102, 173, 223, 239,249
Gilbert, Sgt-Maj. Frank 143
Gill, Ernest 198
Gilson, Robert 79n, 157
Graves, Robert 169
Gray, Charles 236
Gray, Hilda 143

Great Durnford 146
Greenhill House 239, **240**
Greenhill, Job 107–8
Greenslade, Cyrus 28
Grenfell, Capt. Francis 244
Groom, W. H. A. 204, 262
Gurney, Ivor 157, 218

Hacker, William 101
Haldane, Richard 18, 78
Hamilton Battery 80, 207
Hamilton, Lt-Gen. Sir Ian 17, 205, 242
Harcourt, Lewis 61
Harding & Sons 169
Hardy, Thomas 155
Harnham 54
Harris, C. & T. 107
Harrods 114, 116, 196
Hawley, Lt-Col William 214
Henderson, A. E. 246
Henderson, Brig.-Gen. David **93**
Hennessey, Lt Charles 45, 130
Hewetson, Maj. Alexander 91
Heytesbury House 160, 207
Highworth 19
hill badges 70, 73, **73**, 103, 149, 162, 192, 209, 235, 261–3, **261**
Homke, Otto 107
Hopsdorf, Joe **106**
Hornby, Capt. 244n
Hornsby tractor **85**, 86, 148
Horrigan, Gnr John 118
horse racing 141
hospitals 57, **57**, 79, 84, 101, 108, 111, 133–5, **134**, 160, 178, 190, 192, 201, 209, 210, 218, 240, 240–1, 242
House, H. W. 119, 220, 248–9
Hughes, Maj. Christopher 28–31, 195, 237
Hughes-Hallett, Norton 190–1, 244

Idmiston 224, 226
Imber 62, 104, 256
Ingouville-Williams, Maj.-Gen. Edward 239 & n
internees 104–8
invasion fears 16, 83, 245
Isler, Maj. J. L. 105

Jackson, Sir John 49, 53, 80,191, 228, 237
Jaegster, Hubert 108
James, N .D.G. 5, 153
Japanese soldiers 17, 73, 239
Jeeves, Rev L. L. 191
Jenkins, Arthur 156

Keen family 145–6
Kelk, John William 241
Kennet & Avon Canal 104, 196

270

Index

Kermode, Thomas 36, 208
Kinrade, Kim 154
Kipling, Rudyard 178, 212
Kitchener, F. M. Lord 42, 47, 66, 80, 86, 142, 196, 223,
'Kiwi', Bulford 73, **73**, 162, 235, 263
Knook 83, 207
Kohler, Peter 100

Lampen, Capt. F. H. 71
Lansdowne, Marquis of 255
Le Marchant, Sir Gaspard 194
Le Queux, William 16
Lee, John A. 154
Leech, Mrs B. E. 109
Legion of Frontiersmen 138n
Lewis, Gwilym 94
Lighthall, William 16
Lipton's 113
Livermore, Bernard 204
Lomax, Brig.-Gen. Samuel 18
Long Iver 174
Long, Col Walter 250
Longbridge Deverill 228
Longleat 134
Loraine, Capt. Eustace 91
Loraine, Robert 89, 196
Loraine, Winifred 89
Lovatt, Henry 242
Lucy, John F. 24
Ludgershall **23**, 113, 116, 145–6, 166–7, 170, 173, 219–20
Lydeway 63

Mace, Guy 164
Maclaren, Pte Andrew 202
Macmanus, John 178
Maguire, Mrs 167
Marconi, Guglielmo 196–7
Market Lavington 63
Marlborough 150, 169
  College 21n, 28, 155–6, 167, 170, 260
Mary, Queen 25, **43**, 65, 80, 124
Masters, John 154
Maxim, Sir Hiram 61
Maxwell, Capt. W. B. 244
McCracken, Mrs G. M. 201
McCrae, Capt. John 159
McCudden, James 24
McKeon, W. J. 34, 35
Melksham **117**
Mere 118, 133
Merrick, Maj. George **94**
Methuen, F. M. Lord 38, 196
Middle Wallop 119
Military Aeroplane Competition 90–91, **91**
Militia 14 & n, 15, 18
Milne, A. A. 64–5
mobilization 23–5, 246, 248
Monash, Maj.-Gen. John 43, 66

Montgomery, Maj. Bernard 218
Morpurgo, Michael 154
Morris, Frank 156
Morrison, Jean 227
Motor Volunteer Corps 84
Mullens, Maj. R. L. 19
Murray, Ken 46
Mursell, Capt. 107

Neeld, Sir Audrey 255
New Zealand soldiers 26, 42–3, 44–5, 47, 55, 66, 70–3, **72**, 161–2, 166, 169, 181, 192, 193, 213, 233, 234, 249
Newfoundland soldiers 59, 61
novels 153–4
Novogrebelsky, Col 74

Officers' Training Corps 18, 20–21, 21n, 28, 170, 173, 244, 246–9, **247**, 252
Ogden, Arm.-Cpl W. 272
Oxenwood 119
Oxley, Col 248

Paish, F. W. 213
Palmer & Mackay 117
Palmer, Charles 173
Patch, Harry 240
Paterson, Drum Maj. Jack 40
Pearce George 36
Pedler, Bert 230
Pedrail Landship 226–7, **227**
Pembroke, Countess of 134
Pembroke, Earl of 56–7
Penruddocke, Charles 68
Perks, Misses 127
Permeke, Constant 132
Pitcairn Campbell, Lt-Gen. Sir William 158
poets and poetry 154–60
Portuguese labourers 121
postcard publishers 100, 101, 147–8, **148**
postcards 103, 147–50
postmarks 150–2, **152**
prison and internment camps 102–3
prisoners (British) 108–9, **109**
prisoners (German) 104–8, 168, 202, 240–1
Prittie, Henry 22, 153, 162–4
prostitutes 125, 141–6
Pyt House, Tisbury 23, 136, 256

Ramsbury 104
Raven-Hill, Leonard 147
Rawnsley, H. D. 167
recruitment 31–3, 47
Reeves & Son, R. J. 226–7
Rendell, F. 114–5

Reservists 25, 166, 194, 196
Rettberg, Miss 101
Rhymes, John Charles 249
Richardson, A. G. 77
Richardson, Maj. George 70
roads 14, 83–4, 86–7, **87**, 186, 222
Roberts, F. M. Frederick 59, 136n
Robinson, Ira 44
Rocke, Lt. Col. Walter 29, 30
Rockley 38, 39
Roe, F. P. 52
Rollestone Manor 228
Rolls, C. S. 88
Rosener, Frederick 100
Rosenthal, Maj.-Gen. Sir Charles 70
Rothstein, Lce-Cpl Andrew 161
Royal Air Force 202
Royal Army Temperance Association 137, 242
Royal Artillery Harriers 137
Royal Marines 189
Royal Navy 166, 174
Russian soldiers 17, 74
Rutter, Owen 158

Salisbury 24, 48, 116, 120, 138, 140, 144–5, 167, 170
Salisbury, Bishop of 128, 166
Salvation Army 130
Sandhagen, Conrad 107
Sassoon, Siegfried 160
Sayer, Sgt.-Maj. Ernest
Saytch, Alfred 84
Scheumann, Lt Paul 106
Schippers, Joseph 132
Sclater, Gen. Sir Henry 79
searchlights 16
Seely, Col J. E. B. 93–4
Shipton Bellinger 169, 217
shops 115, 233, 240, 242
Shrewton 143, 183
Siepmann, Harry 153, 207
Simonds Brewery 113, 135
Sirdar Rubber Works 117
Smith, Capt. E. L. 199
Smith, Geoffrey 79n
Smith, Victor 44
Smith-Dorrien, Lt-Gen. Sir Horace 90, **182**, 220
Snap 186
Snow, Gnr Ernest 84
Soames, Capt. Arthur 95
Sorley, Charles Hamilton 155
South African soldiers 73, 74, 221
Southwell, Lt Evelyn 38, 158, 254, 260
Spackman, H. & C. 184
Spencer & Co 117
Spicer, Charles 122
'spies' 99–101
Springett, E. J. 135
Spurrier, Steven 196
Stanley, R. 159

271

Index

Stanton 82, 125
Stanton St Bernard 132
Stead, K. R. 178
Stert 102
Stockton 28, 156, 233, 259
Stonehenge 10, 89, 167, 171–3, **172**
Stopford, Sir Frederick 18
Stourton Tower 155
Stratton St Margaret 118
Stratton, Jack 122
Street, A. G. 48, 68, 122
Streets, John 157–8
Suffolk, Lady 132, 134
Sutton Mandeville 261
Sutton Veny 174
Swindon 105, 118, 123, 129, 133, 140–1, 146, 162, 170, 184–5
    GWR works 118
Sykes, Lt-Col William 191
Symons, William 205

tanks 174, 221, 227
Taylor, Pte William **109**
Tedworth Estate 241
    House 241
    Hunt 137
Tennant, Lt Edward 156
Tennant, Norman 228
Territorial Force 18, 22, 173, 200
Thomas, Edward 156, 250
Thomas, Ted 244
Thompson, E. V. 154

Tidworth Military Cemetery 249–50
Tidworth Tattoo 245–6
Tilshead 119
Tisbury 136, 236
Todd, Leon H. 194
Tolkien, J. R. R. 79, 157
Toplis, Percy 181, 245
Tovell, Henri and Tim 211
Trenchard, Maj. Hugh 93
trenches 36–40, **193**, 237
Trinder, A. H. 152
Trowbridge 43, 124, 170
Tytherington 237

United States soldiers 74, 95, 97, 130, 167–8, 201, 202
Upavon 93, **94**
Uphill, Miss 67
Upton Lovel 190
Urchfont 143

van Dyck, P. 132
venereal diseases 132, 145, 178, 187
Voluntary Aid Detachment 133, 262
Volunteer Battalions 14 & n, 18
Volunteer Force 123
Volunteer Training Corps 123, 260

Wadworth's Brewery 113, 194–5
Walker, Matron Jean 241
Warminster 51, 66, 67, 143
Welsh, Hugh 205
West Spring Gun 36, **37**

West, Capt. Graeme 157
Westbury 79 & n, 97
Wheatley Dennis 52, 119, 153, 207
White, Mrs M. B. 133
Williams, Alfred 157, 170
Williams, Ivor 56, 228
Wilson, W. A. 257
Wilson, Staff Sgt Richard 91
Wilton 48, 67, 87, 122
    House 134
Winnie-the-Pooh 64–5, **65**
Winterbourne Stoke 135
wireless telegraphy 16, 18, 19, 20, 89, 196–200, **198**
Women's Emergency Corps 124, 125, 142
Women's Land Corps 124
Woodhenge 199 & n
Woods, Helen B. 131
Woolley & Wallis 169, 194
Wootton Bassett 104
Wroughton 184, 186

Yapp, Sir Arthur 270
Yarnbury Castle 10, 14, 191
Yates, Arthur 113
Yeomanry 14 & n, 18, 222, 254, 255
YMCA 74, 128–30, 137, 190, 204, 217, 220, 231, 235, 239, 261
Young, Francis Brett 8, 78, 154, 234
Young, William Sisley 118
YWCA 129